# FREQUENCY AND RISK ANALYSES IN HYDROLOGY

by

G. W. Kite



WATER RESOURCES PUBLICATIONS
P.O. Box 303
Fort Collins, Colorado 80522, U.S.A.

For Information and correspondence:
WATER RESOURCES PUBLICATIONS
P. O. Box 303
Fort Collins, Colorado 80522

FREQUENCY AND RISK ANALYSES IN HYDROLOGY



ISBN-0-918334-23-3

This publication is printed and bound by LithoCrafters, Chelsea, Michigan, U.S.A.

## Preface

This book is based on an internal report prepared by the author for the Water Resources Branch, Environment Canada. The author is grateful to the Branch Director E. R. Peterson, for permission to use this material.

Table of Contents

## Summary

Every year, floods and droughts cause loss of life and millions of dollars worth of damage around the world. Hydrology is one of the sciences through which man tries to coordinate his requirements with nature's laws in such a way as to minimize these losses.

This report describes some aspects of frequency and risk analysis is they are used in hydrology. The intent of frequency techniques is to derive, from the limited sample data, design events useful in practice. As such, the method is used to approximate, in a mathematically understandable manner, some of the characteristics of the extreme events of the hydrologic cycle. Risk analysis is an aid to assessing the probability that the derived design event will differ from the actual future event.

# CHAPTER 1

## INTRODUCTION

Every year floods take lives and damage properties in many parts of the world. It has been estimated (3) that over the period 1948-1970 floods in Canada have caused an average damage of $5 million per year with the average cost to the Federal Government of $2 million per year. On top of these direct costs are the loss of life, injury, inconvenience and other indirect losses caused by floods.

Forecasts of flood events can generally reduce the damages caused by floods. There are two ways in which this beneficial effect may be achieved. Firstly, and most obviously, warning of an even enables people to evacuate a danger area. If sufficient lead-time is provided, vulnerable possessions may be removed from the danger zone and preparations can be made, such as sandbagging, to minimize property damage.

The second method of achieving benefits through forecasting is to use flood frequency analysis in the design of structures within the flood plain and for flood plain zoning. As examples, a knowledge of magnitude-frequency relationships should be used in the design of dams, highway bridges, railway bridges, culverts, water-supply systems, and flood control structures. The American Water Works Association (1) has reported that out of 293 dam failures in the U. S. and other countries since 1799, about 20% of the failures were due to faulty spillway design. Flood plain zoning ensures that housing and industrial developments are not located in high-risk zones. Or at least, if they are located in such areas there can be no justification for compensation in the inevitable event of flood damage. High-risk zones along the banks of rivers susceptible to floods should be used for activities compatible with this risk such as parkland, recreation areas, some types of crop production or grazing.

Frequency analysis is not only of use as an aid in averting disaster but is also a means of introducing efficient designs. When a hydraulic structure, through inadequate or inaccurate data or methods, is underdesigned, the results are regretably obvious; the dam may fail, the highway may flood or the bridge collapse. This does not happen very often and so the hydrologist, equating nonfailure with success, is satisfied with his design techniques. Nonfailure, however, does not necessarily mean an effecient design. Frequently, structures are overdesigned and hence very safe, but also very expensive. A truly efficient design will be achieved only as the result of studies relating cost to risk and frequency analysis.

The Water Resources Council of the U. S. Government (2) has recently noted that because of the range of uncertainty in design flood analysis there is a need for continued research and development to solve the many unresolved problems. Current methods of providing design floods for hydraulic structures include the deterministic use of meteorological data in techniques such as dynamic flow equations and the so-called Probable Maximum Flood method (PMF), and the stochastic use of frequency analysis techniques. The PMF and similar methods suffer from the major disadvantages of being entirely subjective and of having no associated probability level. This latter is particularly important since to non-technical people it implies that no risk is involved, that the maximum flood cannot exceed this certain limit. This, of course, is untrue and can sometimes have disasterous consequences. Yevjevich (4) has characterised the difference between the PMF method and the frequency analysis method as being between "expediency" and "truth".

Neglecting the PMF approach, this report discusses only the techniques to be used in flood frequency analysis and, to a lesser degree, drought analysis, together with an analysis of the associated risk. All frequency techniques are totally data-dependent. Chapter 2 discusses the data requirements and the basic assumptions involved in the subsequent analyses.

An assumption must be made of a theoretical frequency distribution for the population of events and the statistical parameters of the distribution must be computed from the sample data. Chapters 3-11 describe, for some of the distributions commonly used in hydrology, the form of the distribution, estimation of parameters, estimation of events at given return periods and estimation of confidence limits. The objective of these descriptions is to provide sufficient background information to enable an hydrologist to intelligently select a distribution to use in frequency analysis.

A flood frequency relation for a complete region may be preferable to one developed for a specific site for two reasons: (a) Because of the sample variation possible at a single station, any single station analysis is subject to large error. This error can be reduced by combining data from many sites. (b) There are many more sites where hydrologic data are needed, than there are sites at which data are collected. This means that some form of analysis is required which can transfer data from gauged sites to ungauged sites. Chapter 13 discusses some of the methods of regional analysis presently available.

Both single station and regional frequency analyses involve risks. In determining the design flood for a project, the length of data available, the project life and the allowable probability of failure are all factors to be considered. Chapter 14 reviews the statistical methods available to analyse these risks.

## References


1. American Water Works Association, 1966, Spillway Design Practice, AWWA Manual M13, New York.

2. United States Water Resources Council, 1967, A Uniform Technique for Determining Flood Flow Frequencies, Hydrol. Comm. Bull. No. 15.

3. Water Management Service, 1973, Scope and Plan of Work for Study of Floods and Flood Damages in Canada, unpublished report, Canada Dept. of the Environment, Ottawa.

4. Yevjevich, V., 1972, New Vistas for Flood Investigations, Academia Nationale Dei Lincei, Roma, Quaderno N. 169, pp. 515-546.

# CHAPTER 2

# DATA

## Annual Series and Partial-Duration Series

Starting with the original recorded hydrograph, or with the tabulated data abstracted from the hydrograph, there are two ways in which this data may be used in a frequency analysis. The first method, direct frequency analysis, is to select from the total data only that information which is required in the design process, e.g. maximum instantaneous flow in each year (annual series) all instantaneous flows above a certain base flow (partial-duration), minimum 7-day flow in each year (annual series), etc.. This represents a considerable loss of information, however, since such a few data points are utilised. Dyhr-Nielson (12) has discussed the effects of this loss of information on the estimation of probabilities of extremes.

The second method of using the basic hydrologic data is to design a mathematical model which will describe the observed hydrograph. This model can then be used to generate (simulate) many sets of data from which the required events can be abstracted. As an example, if the objective of the frequency analysis is to define the flow having a magnitude such that it will occur on the average once every 50 years, then 50 years of hydrologic data can be generated and the maximum flow during that period can be found. By generating many different sets of 50-year records, a distribution of 50-year flows can be found and confidence limits can be set on the mean (19). The disadvantage of the data generation method is that it is very unwieldy. It is practicable when dealing with monthly data, less so for daily means and becomes difficult when the need is to generate flows that can be considered instantaneous. In addition, most models cannot successfully generate the extreme peaks and lows of a variable; they only operate well for average conditions.

This report describes only the first method of analysis, direct frequency analysis. Within this method there are two ways in which the required data may be abstracted from the original recorded or tabulated data. As indicated earlier, these are known as annual series and partial-duration series.

An annual series takes one event, and only one event, from each year of the record. A disadvantage of this abstraction technique is that the second or third, etc., highest events in a particular year may be higher than the maximum event in another year and yet they are totally disregarded. This disadvantage is remedied in the partial-duration series method in which all events above a certain base magnitude are included in the analysis. The base is generally selected low enough that at least one event in each year is included. Each event to be included in the partial-duration series must be separate and distinct, i.e. including two consecutive daily flows caused by the same meteorologic event is not valid. If the total number of events which occurred during the entire period of record are ranked without regard to the year in which they occurred and then the n top ranking events are selected, where n is the number of years of record, the events are termed annual exceedences (34).

The recurrence interval of an event of given magnitude is defined as the average length of time between occurrences of that event. This is a probabilistic term and contains no inference of periodicity. The recurrency intervals of annual series and partial duration series have different meanings. In the first case the recurrence interval means the average number of years between the occurrence of an event of a given magnitude as an annual maximum. In the second case the recurrence interval carries no implication of annual maximum.

Chow (8) investigated the theoretical relationship between these two recurrence intervals and their corresponding probabilities. If $P_E$ is the probability of an event in a partial-duration series being equal to or greater than x and if the number of events in the partial-duration series is Nm, where N is the number of years and m is the average number of events per year, then $P_E/m$ is the annual probability of an event being equal to or greater than x. The probability of an event x being the largest of the m events in a year must then be

$$P_M = (1 - P_E/m)^m \tag{2-1}$$

But $P_M$ is then the probability of an annual event of magnitude x and corresponds to the annual series. Substituting the approximation $(1 - P_E/m)^m$ equal to $e^{-P_E}$ and letting $T_E = 1/P_E$ and $T_M = 1/P_M$ where $T_E$ and $T_M$ are the recurrence intervals of the partial-duration and annual series respectively, then

$$T_E = \frac{1}{\ln T_M - \ln (T_M - 1)} \tag{2-2}$$

The following table from Dalrymple (10) compares the recurrence intervals of the two types of series. The difference amounts to about 10% when $P_E$ is 5 years and about 5% when $P_E$ is 10 years. The distinction is only of importance at low recurrence intervals.

To some extent the decision to use an annual series of a partial-duration series depends on the use to which the frequency analysis will be put. Some types of structure, for example erosion protection works, are susceptible not particularly to

Table 2-1

Comparison of Recurrence Intervals for Annual and Partial-Duration Series

| Recurrence intervals (years) | |
|---|---|
| Partial-Duration | Annual Series |
| 0.5 | 1.16 |
| 1 | 1.58 |
| 1.44 | 2 |
| 2 | 2.54 |
| 5 | 5.52 |
| 10 | 10.5 |
| 20 | 20.5 |
| 50 | 50.5 |
| 100 | 100.5 |

one peak flow but more to closely repeated high flows and so a partial-duration series analysis might be more suitable. If the design flood is likely to have a low recurrence interval, then again the partial duration series may be more suitable since the smallest recurrence interval for the annual maximum series in one year.

If flood flows are being investigated, they should preferably be maximum instantaneous flows derived from a continuous hydrograph. Usually, however, instantaneous flows are only available for a comparatively few years, but older data may be available as maximum mean daily flows. In this case, by correlation between the two series, it may be possible to extend the instantaneous maxima back in time (20).

Most data series are incomplete. For various reasons such as mechanical failure, inaccessibility, flood damage to the recorder, etc., some flood peaks are usually missing. This may have a large effect on the recurrence intervals assigned to the floods on record. Dalrymple (10) has described a method by overcoming this problem. By regression with a gauged stream in the same hydrologic region it is possible to estimate the magnitudes of the missing floods at the stream being analysed. As an example of a technique, Langbein (21) has correlated the logarithms of flows standardized on a monthly basis. Dalrymple (10) used the computed flows as an aid in sorting the observed flows by magnitude and assigning recurrence intervals. The computed flows were then discarded and not used in the further analysis.

## Samples From Different Populations

For some stations it may be necessary to carry out more than one frequency analysis on the data. Stoddart and Watt (31) have described how watersheds in southern Ontario have two distinct types of flood; those due to precipitation only, generally occurring in the summer, and those due to snowmelt (sometimes combined with precipitation) generally occurring in the winter or spring. The hydrographs of these two types of flood are quite different and cannot be considered to be from the same population.

Suppose a flood of a given magnitude x would have a recurrence interval of $T_p$ if due to rainfall and $T_S$ is due to snowmelt. The probabilities of not equalling or exceeding x are given as

$$q_p = 1 - 1/T_p \qquad (2\text{-}3)$$

and

$$q_S = 1 = 1/T_S \quad . \qquad (2\text{-}4)$$

So that the probability of not equalling or exceeding x in any year, $q_A$, is given by

$$q_A = q_p \, q_S \qquad (2\text{-}5)$$

and the recurrence interval for the annual flood equalling or exceeding x is

$$T_A = 1/(1 - q_A) \qquad (2\text{-}6)$$

$$T_A = \frac{T_p \, T_S}{(T_p + T_S - 1)} \quad . \qquad (2\text{-}7)$$

The frequency curve of the annual series will be asymptotic to both the rainfall and snowmelt frequency curves. Stoddart and Watt list four possible combinations of these frequency curves.

Various statistical tests are available to test for the presence of data from different distributions. In general, the procedure used in applying these tests is:

(a) State the null hypothesis ($H_o$) that there is no difference between the two data sets.
(b) Choose a statistical test.
(c) Select a significance level ($\alpha$).
(d) Assume the sampling distribution of the statistical test under $H_o$.
(e) On the basis of (b), (c) and (d) define the region of rejection.
(f) Compute the value of the statistical test. If the value is in the region of rejection, reject $H_o$.

There are two types of statistical test, parametric and nonparametric. Parametric tests are associated with certain assumptions about the sample being tested. These assumptions are (Seigel, 27):

(a) The observations must be independent.
(b) The observations must be drawn from normally distributed populations.
(c) These populations must have the same variance.

Generally, non-parametric tests need only assumption (a). In addition, although it is not important here, non-parametric tests do not require as high a level of data measurement as do parametric tests.

The most common parametric test used to test whether or not two samples are from the same population is the t test,

$$t = \frac{|\bar{x}_1 - \bar{x}_2|}{\sigma \sqrt{1/n_1 + 1/n_2}} \qquad (2\text{-}8)$$

where $\bar{x}_1$ and $\bar{x}_2$ are the arithmetic means of the two sub-samples size $n_1$ and $n_2$ and $\sigma$ is the unknown population standard deviation estimated from the sample variances $S_1^2$ and $S_2^2$

$$\sigma = \sqrt{\frac{n_1 S_1^2 + n_2 S_2^2}{n_1 + n_2 - 2}} \qquad (2\text{-}9)$$

If t is greater than the tabulated value of "Students" distribution at significance level $\alpha$ and $n_1 + n_2 - 2$ degrees of freedom and if assumptions (a)-(c) are correct, the samples are not from the same population. However, the applicability of assumptions (b) and (c) must be in doubt; extreme events are not likely to be normally distributed and there is no reason why the variance of events from, say, snowmelt should be the same as the variance from thunderstorm events. For these reasons the t test is not commonly applied in frequency analysis.

Turning to non-parametric tests, amongst those used are the Terry test (33), the Mann and Whitney (24) U test, the Kruskal-Wallis (27) test and the Kolmogorov-Smirnov test (27). To give an example of the use of each of these tests, consider the annual maximum daily discharges of the St. Mary's River at Stillwater, Nova Scotia, for the period 1916 to 1974, listed in Table 2-2. These data will be used in later chapters to test various frequency distributions so it is as well to check their consistency and to see if all the data are from the same population.

Looking at the column of maximum daily discharges in Table 2-2 it is apparent that the floods occur in all months of the year but that November and April are the most popular months; perhaps the November floods are rainfall and the April floods are snowmelt. Figure 2-1 shows the histogram of annual maximum daily discharges by month. The abscissa starts with August and ends with July to emphasize the double peaks of the distribution. The null hypothesis, $H_0$, is that the flood peaks in the months March, April and May are from the same population as the flood

Table 2-2

Sample Data - St. Marys River at Stillwater, Nova Scotia, Station No. 01E0001

| YEAR | ANNUAL MAXIMUM DAILY DISCHARGE | ANNUAL MINIMUM DAILY DISCHARGE |
|---|---|---|
| 1915 | --- | 143 CFS ON SEP 21 |
| 1916 | 10400 CFS ON APR 7 | 37.0 CFS ON AUG 26 |
| 1917 | 10700 CFS ON APR 8 | 37.0 CFS ON SEP 17 |
| 1918 | 20100 CFS ON JAN 5 | 89.0 CFS ON AUG 25 |
| 1919 | 8210 CFS ON APR 19 | 109 CFS ON JUL 10 |
| 1920 | 14300 CFS ON MAR 15 | 109 CFS ON JUL 31 |
| 1921 | 8040 CFS ON MAR 12 | 23.0 CFS ON SEP 22 |
| 1922 | 8210 CFS ON JUL 6 | 98.0 CFS ON JUN 19 |
| 1923 | 13900 CFS ON OCT 3 | 71.0 CFS ON MAR 16 |
| 1924 | 8390 CFS ON JUN 17 | 56.0 CFS ON AUG 5 |
| 1925 | 18500 CFS ON APR 1 | 43.0 CFS ON SEP 8 |
| 1926 | 13000 CFS ON DEC 2 | 37.0 CFS ON SEP 20 |
| 1927 | 16400 CFS ON OCT 20 | 237 CFS ON SEP 18 |
| 1928 | 14500 CFS ON DEC 11 | 63.0 CFS ON SEP 12 |
| 1929 | 13000 CFS ON SEP 20 | 79.0 CFS ON JUL 14 |
| 1930 | 17200 CFS ON MAR 13 | 40.5 CFS ON SEP 14 |
| 1931 | 13900 CFS ON APR 3 | 97.0 CFS ON AUG 13 |
| 1932 | 11900 CFS ON DEC 2 | 60.0 CFS ON JUL 21 |
| 1933 | 13600 CFS ON OCT 9 | 56.0 CFS ON AUG 11 |
| 1934 | 12400 CFS ON APR 15 | 16.0 CFS ON SEP 15 |
| 1935 | 18300 CFS ON JAN 12 | 36.8 CFS ON AUG 22 |
| 1936 | 12900 CFS ON MAR 14 | 122 CFS ON AUG 6 |
| 1937 | 18200 CFS ON NOV 21 | 19.9 CFS ON SEP 8 |
| 1938 | 9900 CFS ON NOV 26 | 140 CFS ON JUN 27 |
| 1939 | 10200 CFS ON APR 21 | 23.8 CFS ON SEP 6 |
| 1940 | 9020 CFS ON DEC 10 | 36.8 CFS ON AUG 19 |
| 1941 | 11800 CFS ON APR 29 | 79.0 CFS ON JUL 8 |
| 1942 | 16100 CFS ON OCT 27 | 5.3 CFS ON SEP 9 |
| 1943 | 16900 CFS ON MAY 5 | 107 CFS ON FEB 3 |
| 1944 | 11800 CFS ON NOV 7 | 20.8 CFS ON AUG 25 |
| 1945 | 13900 CFS ON JAN 18 | 45.3 CFS ON AUG 21 |
| 1946 | 12300 CFS ON DEC 23 | 20.8 CFS ON JUL 21 |
| 1947 | 15100 CFS ON MAY 3 | 16.0 CFS ON AUG 24 |
| 1948 | 11900 CFS ON APR 3 | 118 CFS ON AUG 6 |
| 1949 | 11000 CFS ON NOV 27 | 63.0 CFS ON AUG 3 |
| 1950 | 16000 CFS ON NOV 29 | 14.3 CFS ON OCT 12 |
| 1951 | 11600 CFS ON MAY 30 | 74.0 CFS ON AUG 9 |
| 1952 | 19900 CFS ON JAN 24 | 25.7 CFS ON SEP 17 |
| 1953 | 18600 CFS ON FEB 9 | 25.7 CFS ON AUG 4 |
| 1954 | 18000 CFS ON FEB 25 | 40.5 CFS ON SEP 30 |
| 1955 | 13100 CFS ON FEB 13 | 53.0 CFS ON AUG 18 |
| 1956 | 29100 CFS ON JAN 7 | 41.3 CFS ON OCT 6 |
| 1957 | 10300 CFS ON NOV 4 | 35.6 CFS ON SEP 2 |
| 1958 | 12200 CFS ON NOV 30 | 83.0 CFS ON JUL 29 |
| 1959 | 15600 CFS ON APR 5 | 260 CFS ON JAN 16 |
| 1960 | 12700 CFS ON APR 7 | 7.8 CFS ON SEP 12 |
| 1961 | 13100 CFS ON DEC 30 | 19.1 CFS ON AUG 26 |
| 1962 | 19200 CFS ON APR 3 | 192 CFS ON JUL 1 |
| 1963 | 19500 CFS ON MAY 3 | 119 CFS ON AUG 13 |
| 1964 | 23000 CFS ON APR 17 | 270 CFS ON AUG 23 |
| 1965 | 6700 CFS ON APR 18 | 69.3 CFS ON OCT 1 |
| 1966 | 7130 CFS ON APR 2 | 34.0 CFS ON SEP 4 |
| 1967 | 14300 CFS ON NOV 25 | 132 CFS ON JUL 30 |
| 1968 | 20600 CFS ON AUG 31 | 17.9 CFS ON AUG 20 |
| 1969 | 25600 CFS ON NOV 10 | 29.6 CFS ON SEP 5 |
| 1970 | 8180 CFS ON JUL 13 | 175 CFS ON JAN 28 |
| 1971 | 34400 CFS ON AUG 16 | 82.8 CFS ON JUL 14 |
| 1972 | 16100 CFS ON MAY 17 | 227 CFS ON FEB 13 |
| 1973 | 10200 CFS ON APR 30 | 67.9 CFS ON OCT 4 |
| 1974 | 12300 CFS ON DEC 4 | 42.8 CFS ON AUG 18 |

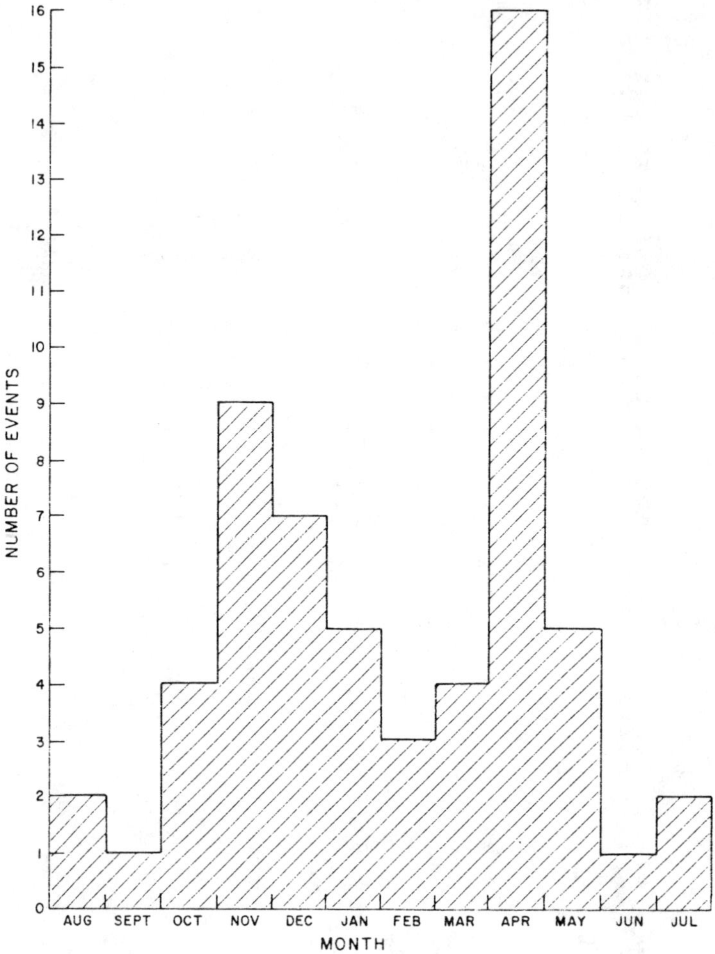


Figure 2-1. St. Marys River at Stillwater Monthly Distribution of Annual Maximum Daily Discharge 1916-1974.

peaks in the months June to February and the level of significance, $\alpha$, is 5%.

*Terry Test* - The 59 annual flood events are ranked in increasing order with the 1965 event of 6,700 cfs ranked 1 and the 1971 event of 34,400 cfs ranked 59. The events are then divided into two samples according to date of occurrence, 25 events in March-May and 34 events in the second sample, June-February. For each event in the first sample the expected value of that order statistic in a sample of 59 from a standardized normal population, $E(Z_{Ni})$, is found from Harter's tables (17). For example, for the 56th rank item in a sample of 59 the expected value is -1.54889 and for the 1st rank item the expected value is 2.31288. The sum of the expected values for each of the 25 ranks in the first subsample is denoted by c and the test statistic is

$$t = \frac{c}{S} \tag{2-10}$$

where S is the standard deviation of c defined as

$$S = \sqrt{\frac{mn}{N(N-1)} \sum_{i=1}^{N} [E(Z_{Ni})]^2} \tag{2-11}$$

where m and n are the sub-sample sizes, N is m+n, the total sample size, and $E(Z_{Ni})$ is the expected ith order statistic from a sample size N from a standardized normal population. For the St. Mary's River data, N = 59, m = 25, n = 34, c = 5.4074 and S = 3.7423 so that t = 1.44. The sampling distribution of t under $H_o$ is normal (0,1) so that the critical value of t in a two-tail test at 5% and N-2 degrees of freedom is 1.96. Since t calculated is less than t critical then $H_o$ cannot be rejected.

*Mann and Whitney Test* - The Mann and Whitney U test (24) is one of the most powerful non-parametric tests and is frequently used (27) as an alternative to the parametric t test. The sample size N is ranked in increasing order as before and the two statistics

$$U_1 = mn + \frac{m}{2}(m + 1) - R_m \tag{2-12}$$

and

$$U_2 = mn - U_1 \tag{2-13}$$

are calculated where m and n are the subsample sizes and $R_m$ is the sums of the ranks assigned to the sample of size m. The smaller of $U_1$ and $U_2$, say U, is then used to calculate the test statistic, Z, as

$$Z = \frac{U - \frac{mn}{2}}{\sqrt{\frac{1}{12}\left[mn\,(m + n + 1)\right]}} \tag{2-14}$$

The Mann and Whitney test is based on a continuous distribution and if tied observations are present a correction must be made. If t is the number of observations tied for a given rank and

$$T = \frac{1}{12}\,(t^3 - t) \tag{2-15}$$

then the test statistic is

$$Z = \frac{U - \frac{mn}{2}}{\sqrt{\frac{mn}{N(N + 1)}\left(\frac{N^3 - N}{12} - \Sigma T\right)}} \tag{2-16}$$

where $\Sigma T$ is the sum of T's over all groups of tied observations. The T's are computed only for those tied observations which contain observations in both samples. For the sample data, $U_1 = 479$, $U_2 = 347$, $\Sigma T = 3.0$ and $Z = 0.83$. For subsamples both greater than 20 items the sampling distribution of Z under $H_0$ is normal (0,1). The two-tail probability of Z may be obtained from tables of the area under the normal curve as 0.39. Since this is greater than $\alpha$ the null hypothesis cannot be rejected.

*Kruskal-Wallis Test* - The statistic used in the Kruskal-Wallis test (27) is H, where

$$H = \frac{12}{N(N + 1)} \sum_{j=1}^{k} \frac{R_j^2}{n_j} - 3\,(N + 1) \tag{2-17}$$

k is the number of samples (2 in this case), $R_j$ is the sum of the ranks in the jth sample, $n_j$ is the number of observations in the jth sample and $N = \sum_{j=1}^{k} n_j$, the total number of observations. For all $n_j > 5$, H is distributed as chi-square with k-1 degrees of freedom. Ties are corrected for by dividing H by

$$1 - \frac{\Sigma T}{N^3 - N} \tag{2-18}$$

where $T = t^3 - t$ and t is the number of observations tied for a given rank. For the sample data H (corrected) is 1.25. From tables of chi-square the critical value at (0.05, 1) is 3.84.

Since the calculated value of H is less than the critical value, the null hypothesis cannot be rejected.

*Kolmogorov-Smirnov* - If two subsamples are from the same population, then the cumulative probability distributions of both subsamples should be fairly close to each other. If the two cumulative distributions are "too far apart" at any point, then the subsamples may be from different populations. Thus a large enough deviation between the two cumulative distributions is evidence for rejecting $H_o$. Using the same class intervals, a cumulative probability distribution is prepared for each subsample and the largest difference in cumulative functions is found

$$D = \max |S_1(x) - S_2(x)| \qquad (2\text{-}19)$$

for a two-tail test. For small sample sizes (<40) in which m=n, Siegel (27) gives tables of critical D. For cases in which m≠n D follows a chi-square distribution with 2 degrees of freedom. For the sample data, using 14 class intervals, D = 0.118. The value of chi-square is then computed as

$$\chi^2 = \frac{4\,D^2\,m\,n}{N} \qquad (2\text{-}20)$$

which for the sample data is 0.80. From tables of chi-square the probability associated with this value at 2 degrees of freedom is 0.75 which is greater than the $\alpha$ of 0.05 and therefore $H_o$ may not be rejected.

In summary, all four non-parametric tests have shown that the difference between the two subsamples, March-May and June-February, are not large enough to reject the hypothesis of a single population.

## Use of Historic Data

Properly authenticated historic events, antedating periods of consecutive records, can be used in frequency analysis to increase the accuracy of the analysis. Discharge estimated on the basis of authenticated stages of historic floods that occurred prior to the modern continuous stage records may be used in conjunction with the modern records to obtain a more accurate probability curve (32).

Benson (3) studying the Susquehanna River at Harrisburg, Pa., found 7 historic floods with stages greater than 18.0 feet in the period 1786 to 1873 and a record of continuous annual maxima ranging from 14.3 to 30.3 feet for the period 1874 to 1947. Since the period of historical floods may have contained an unknown number of events of less than 18.0 feet stage, the problem was to combine the two types of record in the proper proportions to obtain an array of events properly representative of the total period. Benson arrayed all events, historic and recent, in order of descending magnitude using the plotting position m/(n+1). The order number of those flood peaks which were lower than the lowest historical event (the base event, 18.0 feet on the Susquehanna) were then adjusted as

$$m_c = t_b + \frac{n - t_b}{t - t_b} (m - t_b) \qquad (2\text{-}21)$$

where $m_c$ is the corrected order number, m is the original order number, n is the period in years from the first historical event to the most recent recorded event, $t_b$ is the number of events equalling or exceeding the base event, and t is the total number of events.

If a single historic event is known and there is not much difference in magnitude between that event and the highest event in the recent record, then the Equation 2.21 can be used with confidence. If there is a large gap in magnitude then there should be some reasonable certainty that there were few or no events during the ungauged interval which exceeded the highest in the recent period. With a large gap in magnitude and no intimation of what may have happened in between, then Equation 2.21 is no longer applicable. Dalrymple (9) has included further examples of the application of Benson's (3) method.

Glos and Krause (14) have described the augmentation of recorded data with historical data for the Rivers Dnepr in the USSR, Elbe in Czechoslovakia, Main in West Germany and the Spree and Werra in East Germany. As an example, the Elbe River at Decin, Czechoslovakia, has a record of annual floodpeaks for 111 years (1851 to 1962). For purposes of flood frequency analysis historical data can be used to extend this record to 530 years (back to 1432). In this case only the relative magnitudes of the historical floods were used since obviously the absolute magnitudes were not measured.

Glos and Krause used the historical data to increase the accuracy of estimation of the sample mean, $\bar{x}$, and variance; $S^2$, as;

$$\bar{x} = \Sigma x_i p_i \qquad (2\text{-}22)$$

$$S^2 = \Sigma\ (x_i - \bar{x})^2\ p_i \qquad (2\text{-}23)$$

where $p_i$ is a weight given to each event, $\Sigma\ p_i = 1$. For historical floods the weights were assigned as:

$$p = 1/N \qquad (2\text{-}24)$$

where N is the length of the historical period. The events of the annual record were weighted as:

$$p = \frac{N - k}{N\ n} \tag{2-25}$$

where k is the number of historical floods and n is the number of years in the annual record. The computed mean and standard deviation were then used in the standard frequency equation:

$$x_T = \bar{x} + K\ S \tag{2-26}$$

where $x_T$ is the event magnitude at return period T and K, the frequency factor, depends upon the return period required and the distribution characteristics.

Leese (22) has shown how historical flood marks may be used to achieve greater precision in the estimates of flood events and hence greater efficiency in design. The Type I extremal distribution (see Chapter 8 for description) was used on the 29-year record of annual maximum discharges of the River Avon at Bath to estimate floods with return periods of 10, 25, 50, 100 and 1000 years. The procedure was then repeated adding to the data 13 historic floods determined from old flood marks. It was found that the sampling error of the flood estimates was reduced in the second case by between 8% for the 10-year flood and 18% for the 1000-year flood. By applying expressions for the benefits and costs of an imaginary structure Leese found that if the structure were designed on the basis of a 50-year flood then the incorporation of the historic data would reduce the cost of the structure by approximately 1%. More importantly it was found that to obtain the same cost reduction from a continuous record would require a further 20 years of streamflow data.

The use of historical records is apparently quite common in the People's Republic of China. In a recent paper (7) Chen et al. describe some of the advantages of using old records in the estimation of spillway design floods. In historical times the local people would put up monuments marking extreme water levels; for example, along the Chang Jiang or Yangtse River there are a large number of stone sculptures and monuments with inscriptions of the great floods occurring in 1153, 1227, 1560, 1788, 1796, 1860 and 1870. China has an extremely long written history with some flood descriptions being up to 2000 years old. During the Ming and Qing dynasties (a period of nearly 600 years) flood records are complete and in detail. As a further example, the manuscript Shui Jing Zhu (completed in 527 AD) records the level of the flood of 223 AD on the Yi He River. Since the channel is in rock, Chen (7) was able to compute the corresponding discharge from the modern rating curve. The authors point out the necessity of accounting for changes in the local climate and in the physical conditions of the watersheds.

## Errors

Three assumptions are implicit in any frequency analysis (29):

(a) That the data to be analysed describe random events.
(b) That the natural processes involved are stationary with respect to time.
(c) That the population parameters can be estimated from the sample.

This chapter has already discussed the implications of assumption (a) on data selection, emphasizing the fact that all events used must be independent. Assumption (b) is more difficult to guarantee. The earth is in a constant state of flux with innumerable processes affecting the hydrologic cycle and its various components. Statistical tests are available (36) to check for stationarity of time series. Non-stationarity in hydrologic time series is generally due to one of two basic causes:

(a) A slow change in hydrologic parameters such as might be caused by the gradual urbanization of watersheds or (on a different time scale) long-term changes in temperature or precipitation distribution.
(b) Rapid change in parameters caused by, for example, earthquakes, landslides, building of dams.

In a recent paper, Yevjevich (37) gave a plot of annual maximum flows of the Danube River at Orshova in Romania which shows a pronounced upward trend. This trend in discharge is mainly due to the construction of flood protection levels along the Danube and its major tributaries. Non-homogeneities such as this shift the mean value of the distribution and increase the variance.

Another source of error is basic inconsistency in the data due to systematic measurement and computational errors. This subject has been discussed in detail by Dickinson (11), Robertson (24), and Herschy (18), and the particular errors involved in determination of winter flows in Canada have been discussed by Rosenberg and Pentland (26). The major emphasis in this report is on flood flows and these are the very events subject to the maximum measurement error. In fact maximum flows are seldom, if ever, measured because of the difficulties involved firstly in predicting the time at which maximum flow will occur, secondly the difficulty in getting to the gauging site at that time and finally the difficulties of actually carrying out the gauging at the high stage and corresponding high velocities. As a result high flows are normally estimated by extrapolation of the rating curve, estimation of mean velocity from an isolated surface velocity measurement, use of the slope-area method (28) or other similar procedures. The resulting estimates of peak discharge contain a high error component which has been estimated by Blench (5) to be at least ±25%.

Yen and Ang (35) have termed these errors "subjective uncertainty" and have shown that if the individual uncertainties are denoted by $\sigma_1$, $\sigma_2$... then the overall subjective uncertainty $\sigma_n$, can be written as:

$$\sigma_n = [\sigma_1{}^2 + \sigma_2{}^2 + \ldots]^{1/2} \tag{2-27}$$

A third assumption made, in any frequency analysis and a very important one, is that the sample data available can provide good estimates of the population parameters. This assumption is necessary so that estimates of population statistics such as mean, variance, skew, etc. may be derived from the sample. Benson (4) used a theoretical frequency curve to obtain 1000 random events. These base data were then divided into shorter records, e.g. 40 records of 25 events each, 20 records of 50 events each, etc. and Benson investigated the variability of the frequency curves of these samples compared to the original theoretical curve. Benson's conclusions provide estimates of the number of events needed in a sample before estimates of magnitudes at various return periods based on sample statistics are comparable to the values computed from population statistics.

## Plotting Position

Consider a series of 10 annual maximum discharges. Suppose that it is desired to plot these annual maxima on a graph in order to better interpret the data, perhaps detect errors, or to get an idea of which probability distribution to use to describe the data. The ordinate of such a graph conventionally contains the event magnitudes either on a linear or logarithmic scale while the abscissa will be some measure of the probability of occurrence of each event or the average time interval between occurrences (since this is simply the inverse of the probability of occurrence). Frequently the scale of the abscissa will be arranged so that events distributed according to a given distribution will plot as a straight line.

The question then arises as to how to derive the probability of occurrence or average return period of each of the set of annual maximum floods. Sorting the annual events in order of magnitude it is apparent that the largest flood occurs once in the 10 years. But is 10 years the true average return period (i.e. the average interval between occurrences) of this flood? We do not know. In the particular sample a flood of this magnitude occurred only once but in other samples of annual maxima of equal length the same magnitude of flood might occur several times or not at all. That is, the maximum flood in the 10-year sample may have a true return period of 5 years, 50 years, 500 years or any other number of years. And yet, for practical reasons, some probability of occurrence must be assigned to this flood.

For the maximum of a series of 10 annual maximum values it can be shown (34) that the true return period, T, has a 10% probability of being as low as 5 years and a 10% probability of being as high as 100 years. In theory, if the basic data are truly representative, a flood having a return period of 10 years will be equalled or exceeded in a great length of time, on an average of once in 10 years. In 1000 years of record there would be 1000/10, or 100 such floods. However, if these 1000 years were divided into 100 periods of 10 years each, about 37% of such periods would not experience a flood of that magnitude; about 37% of the periods would experience 1 such flood; about 18% would experience 2; about 6% would experience 3; about 1.5% would experience 4; and about 0.5% would experience 5 or more such floods.

As an illustration of the problem consider a large rain storm occurring simultaneously over several adjacent streamflow basins. Imagine that one stream has been gauged for 5 years, another for 10 years and another for 15 years. If the resulting flood runoff is the maximum in the 15 years, it will have an apparent return period of 5 years at one gauge, 10 years at another and 15 years at the third. Which is correct? Obviously the estimate based on 15 years is more correct than those based on 10 or 5 years. In the absence of infinite records some method of correcting for this variation in apparent return period with available record must be derived. The problem of plotting position is to locate the "correct" return period for each event so that the ranges of variation can be replaced by point positions.

The large variation possible in return period, T, can be shown more formally by considering T as an independent random variable which can take the values 1, 2, ... t-1, t, t+1...∞ where t is the number of years from the occurrence of one event until the occurrence of the next event. The distribution of T is then of the form:

$$P(T = t) = p(1 - p)^{t-1} \qquad (2\text{-}28)$$

where p is the probability of the event occuring at any one time. the average value of T or return period is then given by:

$$E(T) = \sum_{t=1}^{\infty} t\, P(T = t) = 1/p \qquad (2\text{-}29)$$

The variance of T can similarly be expressed as:

$$\text{var}\, T = E(T - E(T))^2 \qquad (2\text{-}30)$$

Following Lloyd (23) this expression reduces to:

$$\text{var } T = (1 - p)/p^2 \qquad (2\text{-}31)$$

For small p the variance of the return period can be approximated as 1/p i.e. the same as the expected value. This explains the large observed variations in values of T. Lloyd has also shown that for non-independent sequences of events the variance of the return period is even larger.

The problem of apparent return period varying with sample length is approached in practice by defining a plotting position for the frequency of occurrence, p. The plotting position is generally based on some assumption of the position of the sample estimate of the frequency within a population distribution of frequencies. A few general requirements of any plotting position have been given by Gumbel (15) as:

(a) The plotting position should be such that all observations can be plotted.
(b) The plotting positions should lie between the observed frequencies $(m - 1)/n$ and $m/n$ and should be distribution free (m is the order of the particular event in the series of n maximum events. For the largest of the n events, $m = 1$).
(c) The return period of a value equal to or larger than the largest observation should approach n, the number of observations.
(d) The observations should be equally spaced on the frequency scale, i.e. the difference between the plotting positions of the $(m + 1)$th and the mth observations should be a function of n only and be independent of m.
(e) The plotting position should have an intuitive meaning and should be analytically simple.

*Plotting Position as an Extreme* - The simplest assumption regarding the sample frequency in its population is that the events correspond directly to their observed frequency i.e. the maximum event in a series of 10 independent maxima would have a return period of 10 and a frequency of occurrence of 0.1. This is known as the California method (23) and is given by the general equation

$$p = \frac{m}{n} \qquad (2\text{-}32)$$

That is, considering the frequency interval 0.1 - 0, the California methods uses a plotting position at the upper extreme of this interval. By first considering the observed maxima arranged in a decreasing order of magnitude the observed frequency is equally ligitimately given by

$$p = \frac{m-1}{n} \qquad (2\text{-}33)$$

which corresponds to the lower extreme of the frequency interval noted above. Since the frequencies zero and unity do not exist for an unlimited variate the largest observation of the series cannot be plotted using the function (m-1)/n and the smallest observation cannot be plotted using the function m/n. These two functions therefore fail Gumbel's conditions and are not acceptable as plotting positions unless an upper or lower limit to the population can be envisaged.

*Plotting Position as the Mean* - Foster (13) has pointed out that there is no reasonable basis for assuming that the maximum event in a sample represents exactly the simple (California) frequency. Instead he contended that this maximum event is representative of the whole class of possible events occurring with frequencies less than that of the maximum in the sample, i.e. for a sample of 10 the maximum event represents the interval 0.10 to 0, and therefore should be plotted at the mean of this class interval (i.e. at 0.05 in our example). The Foster (or sometimes called Hazen) plotting position is given by the general equation

$$p = \frac{2m - 1}{2n} \tag{2-35}$$

Equation 2-35 is a compromise between Equations 2-32 and 2-33. When m = 1, the largest event of the sample, this plotting position claims that an event which has already happened once in n years will occur, in the mean, once in 2n years.

Similarly if the observed frequency of an event is assumed to be the mean of the population of frequencies for that event, $\bar{p}$, then

$$p = \frac{\int_0^1 (1-Z^{1/n})\, dZ}{\int_0^1 dZ} \tag{2-36}$$

where Z is the frequency of occurrence of the event in the n year period. For the maximum value, $\bar{p}$ = 1/n+1 so that for a 10-year sample the maximum value would have an assigned average return period of 11 years. Use of the mean frequency leads to the general equation

$$p = \frac{m}{n+1} \tag{2-37}$$

where m is the order of the flood, m being 1 for the largest and n for the smallest event in the n years of record. The probability p is thus the average of the probabilities of all events with rank m in a series of periods each of n years.

*Plotting Position as the Mode* - Another assumption possible is that the observed frequency is the mode (by definition the event which occurs most frequently) of the population of frequencies. Equation 2-28 has no mode and so the type of probability distribution must be incorporated. As an example Equation 2-28 can be used with the Gumbel or type I extremal distribution (15) to yield:

$$Z = (e^{-e^{-y}})^n \qquad (2\text{-}38)$$

where y is a linear function of discharge. For the mode of any distribution the probability density is a maximum i.e. $dZ/dy = 0$ and $d^2Z/dy^2 < 0$, which for equation 2-38 yields

$$p = 1 - (\frac{1}{e})^{1/n} \qquad (2\text{-}39)$$

For the maximum annual event in a 10-year record this equation indicates an average return period of 10-1/2 years. The plotting positions for the maximum events in samples of different size are given in the following table:

Table 2-3

Plotting Positions for Maximum Event Under Modal Assumption

| Sample Size (n) | Plotting Position (p) |
|---|---|
| 2 | 0.393 |
| 5 | 0.181 |
| 10 | 0.0951 |
| 20 | 0.0488 |
| 50 | 0.0198 |
| 100 | 0.00995 |
| 200 | 0.00499 |
| 500 | 0.00199 |
| 1000 | 0.000999 |

It is clear that as the sample size gets larger, p is not significantly different from the simple plotting position $p = m/n$.

Blythe (6) computed the mode of a distribution of extreme values as

$$\text{mode} = \mu - \sigma \frac{\beta_1^{\frac{1}{2}} (\beta_2 + 3)}{2(5\beta_2 - 6\beta_1 - 9)} \qquad (2\text{-}40)$$

where $\mu$, $\sigma$, $\beta_1$ and $\beta_2$ are respectively the mean, standard deviation, coefficient of skew and coefficient of kurtosis. From the normal distribution the probability of occurrence of this mode can be determined. For the maximum values of samples of different sizes Blythe provided the following table. No plotting positions are available by this method for events other than the maximum of the sample, the Blythe recommended that for practical applications for Foster plotting positions (Equation 2-35) should be used.

*Plotting Position as the Median* - A further possible assumption is that at any given order of magnitude within a sample set of events, the return period used should be the median of all possible return periods obtained from a population of equally

Table 2-4

Plotting Positions for Maximum Event of Extremal Distribution

| Sample Size (n) | Plotting Position (p) |
|---|---|
| 2 | 0.306 |
| 5 | 0.143 |
| 10 | 0.077 |
| 20 | 0.041 |
| 60 | 0.014 |
| 100 | 0.0087 |
| 200 | 0.0046 |
| 500 | 0.0018 |
| 1000 | 0.00092 |

sized samples. The return period to be used at any given order of magnitude, j, within a sample of size n events is then given by

$$T = 1/p_j \qquad (2\text{-}41)$$

where $p_j$ is the solution of the binomial equation

$$\sum_{i=0}^{j-1} \binom{n}{i} p_j^i (1 - p_j)^{n-i} = 0.5 \qquad (2\text{-}42)$$

that is, $p_j$ is a solution of a polynomial of order equal to the sample size n and having a number of terms equal to the numerical value of j.

In effect, this method results in a return period of approximately 1.44n for the largest flood in a period of n years. This occurs because, for example, the event with a return period of 144 years has an equal chance of being exceeded or not exceeded in any period of 100 years.

As an example of the use of the median assumption in assigning plotting positions, the maximum event in a sample of 10 maxima would have computed return period, t, of 14.92 years. Beard (2) gave the plotting positions for a sequence of 10 maxima as in Table 2-5.

Table 2-5

Plotting Positions Under Median Assumption

| Magnitude Order (m) | Plotting Position (p) |
|---|---|
| 1 | 0.067 |
| 2 | 0.164 |
| 3 | 0.258 |
| 4 | 0.355 |
| 5 | 0.452 |
| 6 | 0.548 |
| 7 | 0.645 |
| 8 | 0.742 |
| 9 | 0.836 |
| 10 | 0.933 |

The theoretical values of the plotting positions are tedious to compute requiring the solution of a polynomial of degree equal to the sample size and Beard has recommended that the Foster equation (Equation 2-35) which can be transformed to

$$p_j = p_1 + (j-1)\cdot(1-2p_1)/n-1 \qquad (2\text{-}43)$$

where $p_1 = 1/2n$ should be used as an approximate solution.

Banerji and Gupta (1) proposed an alternate solution

to Equation 2-43 as

$$p_1 = 1 - 0.5^{1/n} \tag{2-44}$$

In practice the procedure advocated by Banerji and Gupta is to compute $p_1$ from Equation 2-44, compute the increment $(1-2p_1)/n-1$ and then compute $p_j$ for j = 2, 3... n/2 from Equation 2-43. The remaining plotting positions $p_j$, j = n/2, (n/2+1)...n can be calculated as

$$p_{(n+1)/2} = 0.5 \tag{2-45}$$

$$p_{n-j+1} = 1.0 - p_j \tag{2-46}$$

Beard (2) and Hardison and Jennings (16) have demonstrated that for a normal distribution the average exceedence probability $\bar{p}$, for an event with a 10-year return period estimated from a sample of 10 events is 0.1261. As the length of the sample increases so the average exceedence probability decreases until as $n \to \infty$, $\bar{p} \to p$ where p is the simple probability exceedence estimated as p = m/n.

References

1. Banerji, S., and D. K. Gupta, 1967, On a General Theory of Duration Curve and its Application to Evaluate the Plotting Position of Maximum Probably Precipitation or Discharge, Proc. Symposium on Floods and their Computations, Leningrad, pp. 183-193.

2. Beard, L. R., 1943, Statistical Analysis in Hydrology, Trans., ASCE, Vol. 69, No. 8, Pt. 2, pp. 1110-1160.

3. Benson, M. A., 1950, Use of Historical Data in Flood Frequency Analysis, Trans. Am. Geophys. Union, V. 31, pp. 419-424.

4. Benson, M. A., 1960, Characteristics of Frequency Curves Based on a Theoretical 1000 Year Record, USGS Water Supply Paper No. 1543-A, pp. 51-73.

5. Blench, T., 1959, Empirical Methods, Proceedings of Symposium No. 1, Spillway Design Floods, NRC, Ottawa, pp. 36-48.

6. Blythe, R. H., 1943, Discussion of Paper No. 2201, Trans. ASCE, Vol. 69, No. 8, Pt. 2, pp. 1137-1138.

7. Chen, J. Q., Ye, Y-Y., and W-Y. Tan, 1975, The Important Role of Historical Flood Data in the Estimation of Spillway Design Floods, Scientia Sinica, Vol. XXVIII, No. 5, pp. 669-680.

8. Chow, V. T., 1964, Handbook of Hydrology, McGraw-Hill.

9. Dalrymple, T., 1956, Measuring Floods, Proceedings of IASH Symposia, Darcy, Dijon, Vol. 3, pp. 380-404.

10. Dalrymple, T., 1960, Flood Frequency Analyses, USGS Water Supply Paper No. 1543-A, pp. 1-47.

11. Dickinson, W. T., 1967, Accuracy of Discharge Determinations, Hydrology Paper No. 20, Colorado State University, Fort Collins, Colorado.

12. Dyhr-Nielson, M., 1972, Loss of Information by Discretizing Hydrology Series, Hydrology Paper No. 54, Colorado State University, Fort Collins, Colorado.

13. Foster, H. A., 1936, Methods for Estimating Floods, USGS Water Supply Paper 771, Washington, D. C.

14. Glos, E., and R. Krause, 1967, Estimating the Accuracy of Statistical Flood Values by Means of Long-Term Discharge Records and Historical Data, Proc. Leningrad Symposium on Floods and their Computations, pp. 144-151.

15. Gumbel, E. J., 1958, Statistics of Extremes, Columbia University Press.

16. Hardison, C. H., and M. E. Jennings, 1972, Bias in Computed Flood Risk, Proc. ASCE, Vol. 98, No. HY3, pp. 415-427.

17. Harter, H. L., 1961, Expected Values of Normal Order Statistics, Biometrika, Vol. 48, Nos. 1 and 2, pp. 151-165.

18. Herschy, R. W., 1969, The Evaluation of Errors at Flow Measurement Stations, Technical Note No. 11, Water Resources Board, Reading, England.

19. Kite, G. W., and R. L. Pentland, 1971, Data Generating Methods in Hydrology, Technical Bulletin No. 36, Inland Waters Directorate, Department of the Environment, Ottawa.

20. Kite, G. W., 1974, Case Study of Regional Analysis Techniques for Design Flood Estimation, Can. Jour. Earth Sciences, Vol. 11, No. 6, pp. 801-808.

21. Langbein, W. B., 1960, Plotting Positions in Frequency Analysis, USGS Water Supply Paper No. 1543-A, pp. 48-51.

22. Leese, M. N., 1973, The Use of Censored Data in Estimating T-Year Floods, Proceedings of the UNESCO/WMO/IAHS Symposium on the Design of Water Resources Projects with Inadequate Data, Madrid, Vol. 1, pp. 235-247.

23. Lloyd, E. H., 1970, Return Periods in the Presence of Persistence, Journal of Hydrology, Vol. 10, No. 3, pp. 291-298.

24. Mann, H. B., and D. R. Whitney, 1947, On a Test of Whether One of Two Random Variables is Stochastically Larger than the Other, Ann. Math. Stat., Vol. 18, pp. 50-61.

25. Robertson, A.I.G.S., 1966, The Magnitude of Probable Errors in Water Level Determination at a Gauging Station, Technical Note, No. 7, Water Resources Board, Reading, England.

26. Rosenberg, H. B., and R. L. Pentland, 1966, Accuracy of Winter Streamflow Records, Proc. Eastern Snow Conference, Hartford, Connecticut.

27. Siegel, S., 1956, Nonparametric Statistics for the Behavioral Sciences, McGraw-Hill Series in Psychology, New York, 311 p.

28. Smith, A. G., 1974, Peak Flows by the Slope-Area Method, Technical Bulletin No. 79, Inland Waters Directorate, Canada Dept. of the Environment, Ottawa, 31 pp.

29. Spence, E. S., 1973, Theoretical Frequency Distributions for the Analysis of Plains Streamflow, Can. J. Earth Sci., V. 10, pp. 130-139.

30. Stall, J. B., and J. C. Neill, 1961, A Partial Duration Series for Low-Flow Analyses, Journal of Geophysical Research, Vol. 66, No. 12, pp. 4219-4225.

31. Stoddart, R. B. L., and W. E. Watt, 1970, Flood Frequency Prediction for Intermediate Drainage Basins in Southern Ontario, C. E. Research Report No. 66, Queen's University at Kingston.

32. Subcommittee of the Joint Division Committee of Floods, 1953, Review of Flood Frequency Methods, Trans. ASCE, Vol. 118, pp. 1220-1231.

33. Terry, M. E., 1952, Some Rank Order Tests Which are Most Powerful Against Specified Parametric Alternatives, Ann. Math. Stat., Vol. 23, pp. 346-366.

34. USGS, 1936, Floods in the United States, Water Supply Paper No. 771, Washington, D. C.

35. Yen, B. C., and A. H.-S. Ang, 1971, Risk Analysis in Design of Hydraulic Projects, Proceedings of Symposium on Stochastic Hydraulics, Univ. of Pittsburg, pp. 694-709.

36. Yevjevich, V., 1972, Probability and Statistics in Hydrology, Water Resources Publications, Fort Collins, Colorado.

37. Yevjevich, V., 1972, New Vistas for Flood Investigations, Academia Nationale Dei Lincei, Roma, Quaderno N. 169, pp. 515-546.

# CHAPTER 3

## FREQUENCY DISTRIBUTIONS (GENERAL)

### Introduction

One of the most common problems faced in hydrology is the estimation of a design flood or drought from a fairly short record of streamflows. Plotting the magnitude of the measured events (annual maxima for example), some kind of pattern is generally apparent. The question is how to use this pattern to extend the available data and estimate the design event.

If a large number of observed or measured events are available from a period of record at least as long as the return period of the required design event then the problem is simplified. In the extreme if a large enough sample were available (say one million events) then the design event and its confidence interval could be derived directly from the sample data. This amount of data will not be available, however, and so the sample data are generally used to fit a frequency distribution which in turn is used to extrapolate from the recorded events to the extrapolated either graphically or by estimating the parameters of some standard frequency distribution.

Graphical methods have the advantages of simplicity and visual presentation and the fact that no assumption of distribution periods (i.e. around the centre of the probability distribution) while the events it is required to estimate are generally of large return period (i.e. in the tail of the distribution). Many distributions have similar shape in their centres but differ widely in the tails. It is thus possible to fit several distributions to the sample data and end up with several different estimates of the T-year event. Chi-square and similar tests of goodness of fit can be used to choose the distribution which best describes the sample data but this does not overcome the basic problem.

Once a distribution has been chosen then the second source of error becomes apparent. The statistical parameters of the probability distribution must be estimated from the sample data. Since the sample data is subject to error the method of fitting must minimize these errors and must therefore be as efficient as possible.

### Parameter Estimation

There are four parameter estimation techniques in current use:

1. method of moments,
2. method of maximum likelihood,
3. least squares, and
4. graphical.

*Method of Moments* - The method of moments utilizes either the general equation for calculation of the rth moment about the origin of type is made. These advantages are outweighed, however, by the disadvantage that, given twenty engineers to fit a curve through a set of points, it is highly probably that at least twenty different curves would result. In other words the method is highly subjective and is not compatible with the other phases of engineering design. It must be noted, however, that subjective information need not be totally eliminated and that techniques such as Bayesian analysis may permit its use in the statistical procedure (13).

Numerous different probability or frequency distributions have been used in hydrology (1). Discrete distributions such as the binomial and Poisson have been used to define the average intervals between events (5) and to evaluate risks (7). Continuous distributions such as the normal and lognormal have been used for both annual series (6) and partial duration series (1) to define the magnitude of an event corresponding to a given probability of occurrence. The two types of distribution have also been combined (4), (11), (12) to give models of frequency of occurrence and frequency of magnitude of extreme events.

There are two sources of error in using a frequency distribution to estimate event magnitudes. The first source of error is that it is not known which of the many distributions available is the "true" distribution, i.e., which distribution, if any, the events naturally follow. This is important because the sample events available are usually for relatively low return a distribution, p(x):

$$\mu'_r = \int_{-\infty}^{\infty} x^r \; p(x) \; dx \qquad (3\text{-}1)$$

or the corresponding equation for central moments of the distribution

$$\mu_r = \int_{-\infty}^{\infty} (x - \mu'_1)^r \; p(x) \; dx \qquad (3\text{-}2)$$

where $\mu'_1$ is the first moment about the origin. The method of moments then relates the derived moments to the parameters of the distribution.

As an example consider the application of the method of moments to the normal distribution with two parameters, say $\alpha$ and $\beta$, for which:

$$p(x) = \frac{1}{\beta\sqrt{2\pi}} \, e^{-\frac{1}{2}\left(\frac{x-\alpha}{\beta}\right)^2} \qquad (3\text{-}3)$$

so that

$$\mu'_r = \int_{-\infty}^{\infty} x^r . \frac{1}{\beta\sqrt{2\pi}} e^{-\frac{1}{2}(\frac{x - \alpha}{\beta})^2} dz \qquad (3\text{-}4)$$

For the first moment about the origin

$$\mu'_1 = \int_{-\infty}^{\infty} \frac{x}{\beta\sqrt{2\pi}} e^{-\frac{1}{2}(\frac{x - \alpha}{\beta})^2} dz \qquad (3\text{-}5)$$

Substituting z for $(x-\alpha)/\beta$

$$\mu'_1 = \frac{1}{\sqrt{2\pi}} \int_{-\infty}^{\infty} (z\beta + \alpha) e^{-z^2/2} dz \qquad (3\text{-}6)$$

but $ze^{-z^2/2}$ is an odd function and the integral of an odd function between symmetric limits is zero so

$$\mu'_1 = \frac{\alpha}{\sqrt{2\pi}} \int_{-\infty}^{\infty} e^{-z^2/2} dx = \alpha \qquad (3\text{-}7)$$

i.e. for this distribution the parameter $\alpha$ is the arithemetic mean (the first moment about the origin), say $\bar{x}$.

Utilizing Equation 3-2 for the second central moment

$$\mu_2 = \int_{-\infty}^{\infty} (x - \bar{x})^2 \frac{1}{\beta\sqrt{2\pi}} e^{-\frac{1}{2}(\frac{x - \alpha}{\beta})^2} dx \qquad (3\text{-}8)$$

but from 3-7 $\alpha = \bar{x}$. Substituting z for $(x - \bar{x})/\beta$

$$\mu_2 = \frac{\beta^2}{\sqrt{2\pi}} \int_{-\infty}^{\infty} z^2 e^{-z^2/2} dx \qquad (3\text{-}9)$$

Since the function $z^2e^{-z^2/2}$ is even, and substituting y for $z^2/2$

$$\mu_2 = \frac{2\beta^2}{\sqrt{2\pi}} \int_{o}^{\infty} y^{\frac{1}{2}} e^{-y} dy \qquad (3\text{-}10)$$

$$\mu_2 = \frac{2\beta^2}{\sqrt{2\pi}} \frac{\sqrt{\pi}}{2} = \beta^2 \qquad (3\text{-}11)$$

i.e. parameter $\beta$ is the square root of the second central moment.

To compute higher order central moments it is frequently easier to compute the moments about the origin, $\mu'_r$, and then convert to central moments, $\mu_r$, using the expression

$$\mu_r = \sum_{j=0}^{r} \binom{r}{j} \mu'_{r-j} \left(-\mu'_1\right)^j \qquad (3\text{-}12)$$

so that, for example,

$$\mu_2 = \mu'_2 - \mu'^{\,2}_1 \qquad (3\text{-}13)$$

*Method of Maximum Likelihood* - The principle of maximum likelihood states that for a distribution with a probability density function $p(x;\alpha,\beta,...)$ where $\alpha,\beta...$ are the distribution parameters to be estimated, then the probability of obtaining a given value of x, $x_i$, is proportional to $p(x_i;\alpha,\beta,...)$ and the joint probability, L, of obtaining a sample of n values $x_1$, $x_2$,... $x_n$ is proportional to the product

$$L = \prod_{i=1}^{n} p(x_i;\alpha,\beta,...) \qquad (3\text{-}14)$$

This is called the likelihood; the method of maximum likelihood is to estimate, $\alpha$, $\beta$,..., such that L is maximized. This is obtained by partially differentiating L with respect to each of the parameters and equating to zero.

Frequently ln L is used instead of L to simplify computations. As an example of the application of the maximum likelihood technique consider again the normal distribution with parameters $\alpha$ and $\beta$:

$$p(x) = \frac{1}{\beta\sqrt{2\pi}} e^{-\frac{1}{2}\left(\frac{x-\alpha}{\beta}\right)^2} \qquad (3\text{-}15)$$

so that

$$L = \left\{\frac{1}{\beta\sqrt{2\pi}}\right\}^n e^{\frac{\sum_{i=1}^{n}(x_i-\alpha)^2}{2\beta^2}} \qquad (3\text{-}16)$$

Taking logarithms

$$\ln L = -\frac{n}{2}\ln 2\pi - \frac{n}{2}\ln \beta^2 - \frac{\sum_{i=1}^{n}(x_i-\alpha)^2}{2\beta^2} \qquad (3\text{-}17)$$

Differentiating with respect to the parameters $\alpha$ and $\beta^2$ and equating to zero

$$\frac{\partial \ln L}{\partial \alpha} = \frac{\sum_{i=1}^{n} (x_i - \alpha)}{\beta^2} = 0 \qquad (3\text{-}18)$$

so that

$$\sum_{i=1}^{n} x_i - \sum_{i=1}^{n} \alpha = 0 \qquad (3\text{-}19)$$

but

$$\sum_{i=1}^{n} \alpha = n\alpha \qquad (3\text{-}20)$$

so that

$$\alpha = \frac{\sum_{i=1}^{n} x_i}{n} \qquad (3\text{-}21)$$

Also

$$\frac{\partial \ln L}{\partial \beta^2} = -\frac{n}{2\beta^2} + \frac{\sum_{i=1}^{n} (x_i - \alpha)^2}{2\beta^4} = 0 \qquad (3\text{-}22)$$

so that

$$\beta^2 = \frac{\sum_{i=1}^{n} (x_i - \alpha)^2}{n} \qquad (3\text{-}23)$$

i.e. parameters $\alpha$ and $\beta$ are the mean and standard deviation of the distribution.

*Least Squares* - The least squares estimation method (14) consists of fitting a theoretical function to an empirical distribution. The sum of squares of all deviations of observed points from the fitted function is then minimized. Thus to fit a function

$$\hat{y} = p(x;\alpha,\beta,\ldots) \qquad (3\text{-}24)$$

the sum to be minimized is

$$S = \sum_{i=1}^{n} (y_i - \hat{y}_i)^2 \qquad (3\text{-}25)$$

or

$$S = \sum_{i=1}^{n} (y_i - f(x_i;\alpha,\beta,\ldots))^2 \qquad (3\text{-}26)$$

where $x_i$ and $y_i$ are coordinates of observed points, $\alpha$, $\beta$,... are parameters and n is the sample size. To obtain the minimum sum of squares Equation 3-25 is partially differentiated with respect to the parameter estimates a, b,...

$$\frac{\partial \sum_{i=1}^{n} (y_i - \hat{y}_i)^2}{\partial a} = o \qquad (3\text{-}27)$$

$$\frac{\partial \sum_{i=1}^{n} (y_i - \hat{y}_i)^2}{\partial b} = o \qquad (3\text{-}28)$$

These partial derivatives give a number of equations equal to the number of parameters to be estimated. In order that the least squares method be an efficient estimator three conditions must be satisfied:

1. The deviations $y_i - \hat{y}_i$ be normally or at least symmetrically distributed.
2. The population variance of the deviations be independent of the magnitude of $y_i$.
3. The population variance of the deviations along the least squares curve be constant.

*Graphical Method* - The graphical method consists of fitting a function

$$\hat{y} = p(x;\alpha,\beta,...) \qquad (3\text{-}29)$$

visually through the set of coordinate pairs. To estimate m parameters, m points on the curve are selected giving m equations to solve. The process may be simplified by trying various types of graph paper using transformed coordinates until a straight line fit is possible.

In ascending order of efficiency the four methods of estimation may be listed as graphical, least squares, method of moments, maximum likelihood. To offset its great efficiency, however, the method of maximum likelihood is somewhat more difficult to apply.

These four methods will, more or less efficiently, compute the parameters of a distribution from a particular data sample. As discussed earlier this sample may or may not be typical of the underlying population and use of the sample estimates of parameters may bias the results. A method of Bayesian analysis has been recommended (13) which would use distributions of parameter values rather than point values. These distributions might be obtained from regional analysis.

There are also general criteria with which a distribution should comply before being used in hydrology. As an example, since negative flows are inacceptable a distribution should be bounded on the lower tail. This criterion would eliminate both the normal and the double exponential or Gumbel distributions. In fact, both of these distributions are used in hydrology by ignoring negative flows, replacing them with zeros, or treating the probability of zero flow as a probability mass (14).

## Frequency Factor

Having selected a distribution and estimated its parameters the question is how to use this distribution in the frequency analysis. Chow (1) has proposed a general equation:

$$x_T = \mu + K\,\sigma \qquad (3\text{-}30)$$

or

$$x_T = m_1' + K\sqrt{m_2} \qquad (3\text{-}31)$$

where $x_T$ is the event magnitude at a given return period, T, $\mu$ and $\sigma$ are the population mean and standard deviation estimated by sample moments $m_1'$ and $\sqrt{m_2}$ and K is a frequency factor which is a function of the return period and the distribution parameters. For any chosen distribution a relationship can be derived between the return period and the frequency factor.

## Standard Error

A measure of the variability of the resulting event magnitudes is the standard error of estimate. Each method of estimating the parameters of a distribution can also be used to derive the standard error. The standard error of estimate is defined as:

$$S = \left\{\left[\sum_{i=1}^{n} (x_i - \hat{x}_i)^2\right]/n\right\}^{\frac{1}{2}} \qquad (3\text{-}32)$$

where $\hat{x}_i$ is the computed estimate of recorded event $x_i$.

The differences between the recorded and the computed events may have two origins:

1. The choice of a theoretical population distribution from the sample may be wrong, and
2. The errors in the parameters of the chosen population distribution may be inaccurate due to the shortness of the sample data.

The standard error of estimate accounts for only the second of these causes.

*Standard Error by Moments* - In general the variance of the T-year event, $x_T$, can be derived from moment estimates as follows. If $x_T$ depends upon the first three moments of a distribution and the return period, T, as

$$x_T = f(m_1', m_2, m_3, T) \tag{3-33}$$

then, since T is not a variable, the variance of $x_T$ is given by

$$S_T^2 = \left(\frac{\partial x_T}{\partial m_1'}\right)^2 \text{var } m_1' + \left(\frac{\partial x_T}{\partial m_2}\right)^2 \text{var } m_2 + \left(\frac{\partial x_T}{\partial m_3}\right)^2 \text{var } m_3$$

$$+ 2\frac{\partial x_T}{\partial m_1'} \cdot \frac{\partial x_T}{\partial m_2} \text{ cov } (m_1', m_2) + 2\frac{\partial x_T}{\partial m_1'} \circ \frac{\partial x_T}{\partial m_3} \text{ cov } (m_1', m_3)$$

$$+ 2\frac{\partial x_T}{\partial m_2} \circ \frac{\partial x_T}{\partial m_3} \text{ cov } (m_2, m_3) \tag{3-34}$$

The actual relationship between $x_T$ and the moments has been given in 3-30 as

$$x_T = m_1' + K\sqrt{m_2} \tag{3-30}$$

where K is a function of the coefficient of skew, g, and $g = m_3/m_2^{3/2}$ so that:

$$\frac{\partial x_T}{\partial m_1'} = 1 \tag{3-35}$$

$$\frac{\partial x_T}{\partial m_2} = \frac{K}{2\sqrt{m_2}} + \sqrt{m_2}\ \frac{\partial K}{\partial m_2} = \frac{K}{2\sqrt{m_2}} - \frac{3g}{2\sqrt{m_2}} \cdot \frac{\partial K}{\partial g} \tag{3-36}$$

and

$$\frac{\partial x_T}{\partial m_3} = \frac{\partial x_T}{\partial K} \cdot \frac{\partial K}{\partial g} \cdot \frac{\partial g}{\partial m_3} = \frac{1}{m_2} \cdot \frac{\partial K}{\partial g} \tag{3-37}$$

so that

$$S_T^2 = \text{var } m_1' + \left[\frac{1}{2\sqrt{m_2}}\left(K - 3g\,\frac{\partial K}{\partial g}\right)\right]^2 \text{var } m_2 + \frac{1}{m_2^2}\left(\frac{\partial K}{\partial g}\right)^2 \text{var } m_3$$

$$+ \frac{1}{\sqrt{m_2}}\left(K - 3g\,\frac{\partial K}{\partial g}\right) \text{cov}\left(m_1', m_2\right) + \frac{2}{m_2} \cdot \frac{\partial K}{\partial g} \text{ cov}\left(m_1', m_3\right)$$

$$+ \frac{1}{m_2^{3/2}} \cdot \frac{\partial K}{\partial g} \left( K - 3g \frac{\partial K}{\partial g} \right) \text{cov} \left( m_2, m_3 \right) \tag{3-38}$$

but from Kendall and Stuart (8) pp. 230-232

$$\text{var } m_1' = \mu_2/n \tag{3-39}$$

$$\text{var } m_2 = \frac{1}{n} \left\{ \mu_4 - \mu_2^2 \right\} \tag{3-40}$$

$$\text{var } m_3 = \frac{1}{n} \left\{ \mu_6 - \mu_3^2 - 6\mu_4\mu_2 + 9\mu_2^3 \right\} \tag{3-41}$$

$$\text{cov} \left( m_1', m_2 \right) = \mu_3/n \tag{3-42}$$

$$\text{cov} \left( m_1', m_3 \right) = \frac{1}{n} \left\{ \mu_4 - 3\mu_2^2 \right\} \tag{3-43}$$

and

$$\text{cov} \left( m_2 , m_3 \right) = \frac{1}{n} \left\{ \mu_5 - 4\mu_3\mu_2 \right\} \tag{3-44}$$

Substituting these results into Equation 3-38, using population parameters throughout, and simplifying by substituting

$$\gamma_1 = \mu_3/\mu_2^{3/2} \tag{3-45}$$

$$\gamma_2 = \mu_4/\mu_2^2 \tag{3-46}$$

$$\gamma_3 = \mu_5/\mu_2^{5/2} \tag{3-47}$$

and

$$\gamma_4 = \mu_6/\mu_2^3 \tag{3-48}$$

the following expression results:

$$S_T^2 = \frac{\mu_2}{n} \Bigg\{ 1 + K\gamma_1 + \frac{K^2}{4} \left[ \gamma_2 - 1 \right] + \frac{\partial K}{\partial \gamma_1} \Bigg[ 2\gamma_2 - 3\gamma_1^2 - 6 + K \left( \gamma_3 - 6\gamma_1\gamma_2/4 - 10\gamma_1/4 \right) \Bigg] + \left( \frac{\partial K}{\partial \gamma_1} \right)^2 \Bigg[ \gamma_4 - 3\gamma_3\gamma_1 - 6\gamma_2 - 9\gamma_1^2\gamma_2/4 + 35\gamma_1^2/4 + 9 \Bigg] \Bigg\} \tag{3-49}$$

The standard error of estimate, $S_T$, at any given return period, T, can now be found by substituting n, $\mu_2$, $\gamma_1 - \gamma_4$, K and $\partial K/\partial \gamma_1$ for a particular distribution and sample and taking the square root of 3-49.

In later chapters the expression

$$S_T = \delta\sqrt{\frac{\mu_2}{n}} \tag{3-50}$$

is used so that tables of $\delta$ can be prepared independently of $\mu_2$ and n. Note that for a distribution in which the frequency factor, K, is not dependent on the coefficient of skew the expression for $\delta$ simplifies to:

$$\delta = \left\{1 + K\gamma_1 + \frac{K^2}{4}\left[\gamma_2 - 1\right]\right\}^{\frac{1}{2}} \tag{3-51}$$

*Standard Error by Maximum Likelihood* - Assume that for a particular distribution the three parameters $\alpha$, $\beta$ and $\gamma$ have been estimated by the method of maximum likelihood. The T-year event, $x_T$, is then some function of these parameters and the return period T:

$$x_T = f(\alpha, \beta, \gamma, T) \tag{3-52}$$

Since T is not a variable the standard error of $x_T$, $S_T$, may be written as

$$S_T^2 = \left(\frac{\partial x}{\partial \alpha}\right)^2 \text{var } \alpha + \left(\frac{\partial x}{\partial \beta}\right)^2 \text{var } \beta + \left(\frac{\partial x}{\partial \gamma}\right)^2 \text{var } \gamma$$

$$+ 2\frac{\partial x}{\partial \alpha}\frac{\partial x}{\partial \beta} \text{cov } (\alpha,\beta) + 2\frac{\partial x}{\partial \alpha} \cdot \frac{\partial x}{\partial \gamma} \text{cov } (\alpha,\gamma)$$

$$+ 2\frac{\partial x}{\partial \beta}\frac{\partial x}{\partial \gamma} \text{cov } (\beta,\gamma) \tag{3-53}$$

The partial derivatives may be obtained directly from the likelihood equation and Fisher (3), quoted in Kendall and Stuart (9), has shown that the variance/covariance matrix

$$\begin{bmatrix} \text{var } \alpha & \text{cov } (\alpha,\beta) & \text{cov } (\alpha,\gamma) \\ & \text{var } \beta & \text{cov } (\beta,\gamma) \\ & & \text{var } \gamma \end{bmatrix} \tag{3-54}$$

is the inverse of the symmetric matrix

$$\begin{bmatrix} -\dfrac{\partial^2 L}{\partial \alpha^2} & -\dfrac{\partial^2 L}{\partial \alpha \partial \beta} & -\dfrac{\partial^2 L}{\partial \alpha \partial \gamma} \\ & -\dfrac{\partial^2 L}{\partial \beta^2} & -\dfrac{\partial^2 L}{\partial \beta \partial \gamma} \\ & & -\dfrac{\partial^2 L}{\partial \gamma^2} \end{bmatrix} \qquad (3\text{-}55)$$

where L is the likelihood equation for the particular distribution. Analytical expressions for each of the variances and covariances may therefore be derived in terms of the minors and determinant of the symmetric matrix above, as for example:

$$\text{var } \alpha = \left[\frac{\partial^2 L}{\partial \beta^2} \cdot \frac{\partial^2 L}{\partial \gamma^2} - \left\{\frac{\partial^2 L}{\partial \beta \partial \gamma}\right\}^2\right] \Big/ D \qquad (3\text{-}56)$$

where D, the determinant, is given by

$$D = -\frac{\partial^2 L}{\partial \alpha^2}\left[\frac{\partial^2 L}{\partial \beta^2} \cdot \frac{\partial^2 L}{\partial \gamma^2} - \left\{\frac{\partial^2 L}{\partial \beta \partial \gamma}\right\}^2\right] + \frac{\partial^2 L}{\partial \alpha \partial \beta}\left[\frac{\partial^2 L}{\partial \alpha \partial \beta} \cdot \frac{\partial^2 L}{\partial \gamma^2} - \frac{\partial^2 L}{\partial \beta \partial \gamma} \cdot \frac{\partial^2 L}{\partial \alpha \partial \gamma}\right] - \frac{\partial^2 L}{\partial \alpha \partial \gamma}\left[\frac{\partial^2 L}{\partial \alpha \partial \beta} \cdot \frac{\partial^2 L}{\partial \beta \partial \gamma} - \frac{\partial^2 L}{\partial \beta^2}\frac{\partial^2 L}{\partial \alpha \partial \gamma}\right] \qquad (3\text{-}57)$$

Evaluation of the standard error of estimate is then simple.

For distribution such as the lognormal, extremal type III and log-Pearson type III, which are logarithmic transformations of simple distributions there are two possible methods of computing event magnitudes and standard errors of event magnitudes. The first method is to develop analytical relationships for the frequency factor K and the parameter $\delta$ of the transformed distributions and use these with the mean and standard deviation of the original data. The second method is to logarithmically transform the original data and use the mean and standard deviation of the logarithms together with the K and $\delta$ values for the simple untransformed distribution.

If the distribution of the T-year event were known, then confidence limits could be derived for the event. Two methods can be used to find this distribution: analytical and empirical.

The analytical approach utilizes the probability distribution fitted to the observed data. Cramer (2) has shown that for a sample of n values fitted with a distribution having a probability density function p(x) and a cumulative probability function P(x) then the density function, g(x), of a random variable $x_T$ is given by

$$g(x) = \binom{n}{m} (n-m)(P(x))^m (1-P(x))^{n-m-1} p(x) \qquad (3\text{-}58)$$

where m is n P and P is the cumulative probability associated with event magnitude $x_T$. Now if G(x) represents the cumulative probability function of the random variable $x_T$

$$G(x) = \int_0^{x_o} g(x)\, dx \qquad (3\text{-}59)$$

then the upper or lower confidence limits, $x_o$, for the T-year event, $x_T$, may be found by solving Equation 3-59 for different levels of significance. For example, for the 95% upper confidence limits $x_o$ must be found for G(x) = 0.95.

Unfortunately the evaluation of Equation 3-59 for $x_o$ involves a lot of approximations and still must be carried out by a numerical iteration process. Because of these computational difficulties the second method of deriving confidence limits, the empirical method, is frequently used. In the empirical method Equation 3-30 is used to compute the mean T-year event, $x_T$, and the standard error of the T-year event, $S_T$ is computed by either moments or maximum likelihood. The assumption is then made (10) that the distribution of T-year events is normal so that the confidence interval is given by

$$x_T \pm t\, S_T \qquad (3\text{-}60)$$

where t is the standard normal deviate corresponding to the required confidence level.

References

1. Chow, V. T., 1964, Editor-in-Chief, Handbook of Applied Hydrology, McGraw-Hill.

2. Cramer, H., 1946, Mathematical Methods of Statistics, Princeton University Press.

3. Fisher, R. A., 1921, On the Mathematical Foundations of Theoretical Statistics, Phil. Trans. Roy. Soc., Series A, Vol. 222, pp. 309-368.

4. Frost, J., and R. T. Clarke, 1972, Estimating the T-year Flood by the Extension of Records of Partial Duration Series, Bull. IAHS, Vol. XVIII, No. 1, pp. 209-217.

5. Hall, W. A., and D. T. Howell, 1963, Estimating Flood Probabilities within Specific Time Intervals, J. Hydrology, Vol. 1, No. 1, pp. 265-271.

6. Hardison, C. H., 1969, Accuracy of Streamflow Characteristics, USGS Professional Paper No. 650-D, pp. D210-D214.

7. Kalinin, G. P., 1960, Calculation and Forecasts of Streamflow from Scanty Hydrometric Readings, Trans. Interregional Seminar on Hydrologic Networks and Methods, Bangkok, 1959, WMO Flood Control Series No. 15, pp. 42-52.

8. Kendall, M. G., and A. Stuart, 1963, The Advanced Theory of Statistics, Vol. I, Distribution Theory, Griffin, London.

9. Kendall, M. G., and A. Stuart, 1967, The Advanced Theory of Statistics, Vol. II, Inference and Relationship, Griffin, London.

10. Kite, G. W., 1975, Confidence Limits for Design Events, Wat. Res. Res., Vol. 11, No. 1, pp. 48-53.

11. Shane, R. M., 1966, A Statistical Analysis of Base-Flow Flood Discharge Data, Cornell University, PhD Thesis.

12. Todorovic, P., and D. A. Woolhiser, 1972, On the Time When the Extreme Flood Occurs, Wat. Res. Res., Vol. 8, No. 6, pp. 1433-1438.

13. Wood, E. F., and I. Rodriquez-Iturbe, 1975, Bayesian Inference and Decision Making for Extreme Hydrologic Events, Wat. Res. Res., Vol. 11, No. 4, pp. 533-542.

14. Yevjevich, V., 1972, Probability and Statistics in Hydrology, Water Resources Publications, Fort Collins, Colorado.

## CHAPTER 4

## DISCRETE DISTRIBUTIONS

### Binomial

Tossing a coin or drawing a card from a pack are examples of a Bernoulli trial. Bernoulli trials operate under three conditions:

1. Any trial can have only one ot two possible outcomes; success or failure, true or false, rain or no rain, etc.
2. Successive trials are independent.
3. Probabilities are stable.

Under these conditions the probability of x successes in n trials is given by the binomial distribution as

$$p(x) = \binom{n}{x} p^x q^{n-x} \qquad (4\text{-}1)$$

where $\binom{n}{x}$, sometimes written as nCx of $C^n_x$, is the number of combinations of n events taken x at a time,

$$\binom{n}{x} = \frac{n!}{x!\ (n-x)!} \qquad (4\text{-}2)$$

p is the probability of occurrence of an event, for example the probability of success in tossing a coin, q is the probability of failure,

$$q = 1-p \qquad (4\text{-}3)$$

and x is the variate or the number of successful trials.

As an example of the use of the binomial distribution suppose that a dam has a projected life of 50 years and we wish to evaluate the probability that a flood with a return period of 100 years will occur once during the life of the dam. Then $p = 1/T = 0.01$, $q = 1-p = 0.99$, $x = 1$ and $n = 50$, so that

$$p(1) = \binom{50}{1} (0.1)^1 (.99)^{49} = 0.306 \qquad (4\text{-}4)$$

i.e., there is about a 31% chance that an event of that magnitude will occur once in the life of the dam.

### Poisson

The terms of a binomial expansion are a little inconvenient to compute in any large number. Provided that p is small (say

$< 0.1$) and n is large (say $n > 30$) and the mean n p is constant well defined, it can be shown (3) that as $p \to 0$, $q \to 1$ and $n \to \infty$.

$$(p+q)^n \to e^{-\lambda} \cdot e^{\lambda} = e^{-\lambda} + \lambda e^{-\lambda} + \frac{\lambda^2 e^{-\lambda}}{2!} + \ldots\ldots \qquad (4\text{-}5)$$

This is known as the Poisson expansion and is generally written

$$p(x) = \frac{\lambda^x e^{-\lambda}}{x!} \qquad (4\text{-}6)$$

where $\lambda = n\,p$ is the mean. The finite binomial distribution can thus be approximated by the infinite Poisson distribution provided that the following four conditions apply:

1. The number of events is discrete.
2. Two events cannot coincide.
3. The mean number of events in unit time is constant.
4. Events are independent.

Repeating the previous example, the probability that a 100 year return period flood will occur once in a 50 year period is seen to be

$$p(1) = \frac{0.5^1 e^{-0.5}}{1.} = 0.303 \qquad (4\text{-}7)$$

which agrees well with the result obtained from the binomial expansion.

Equation 4-6 will give not only the probability of one event occurring in a given time but also the probability that two events may occur in that time, that three may occur, etc., etc. The probability that one or more events occur will therefore be given as a summation of Equation 4-6.

$$P(1,2\ldots\infty) = \sum_{x=1}^{\infty} p(x) \qquad (4\text{-}8)$$

but

$$\sum_{x=1}^{\infty} p(x) = \sum_{x=0}^{\infty} p(x) - p(0) \qquad (4\text{-}9)$$

which, from Equation 4-5, gives

$$P(1,2\ldots\infty) = e^{-\lambda}e^{\lambda} - e^{-\lambda} \qquad (4\text{-}10)$$

or

$$P(1,2\ldots\infty) = 1 - e^{-\lambda} \qquad (4\text{-}11)$$

Table 4-1 shows, for $\lambda = 0.5$, the values of $P(x)$ at $x = 1, 2, 3, 4$ and at $x = \infty$.

Table 4-1

Some Values of Probability of One or More Events for a Poisson Distribution with $\lambda = 0.5$

| Variate Value, x | Probability P(1...x) |
|---|---|
| 1 | 0.30326 |
| 2 | 0.37908 |
| 3 | 0.39171 |
| 4 | 0.39329 |
| : | : |
| ∞ | 0.39347 |

Abbreviating the probability of one or more occurrences $P(1,2,\ldots\infty)$ to P and replacing $\lambda$, the average number of events per time period, by $\Delta t/T$ where $\Delta t$ is the time interval being considered (e.g. project life) and T is the event return period, then Equation 4-11 becomes:

$$P = 1 - e^{-\Delta t/T} \qquad (4\text{-}12)$$

Hall and Howell (1) have prepared a table (Table 4-2) showing values of P for different time intervals and return periods.

Table 4-2

Probabilities of One or More Occurrences of Events With Different Return Periods in Different Time Intervals

| Time Interval Δt | Return Period T | | | | | | | | | | |
|---|---|---|---|---|---|---|---|---|---|---|---|
| | .1 | .2 | .4 | 1 | 2 | 4 | 10 | 20 | 40 | 100 | 200 |
| .1 | .632 | .393 | .221 | .095 | .049 | .025 | .010 | .005 | .003 | .001 | .0005 |
| .2 | .865 | .632 | .393 | .181 | .095 | .049 | .020 | .010 | .005 | .002 | .001 |
| .4 | .982 | .865 | .632 | .380 | .181 | .095 | .039 | .020 | .010 | .004 | .002 |
| 1 | 1.000 | .993 | .918 | .632 | .393 | .221 | .095 | .049 | .025 | .010 | .005 |
| 2 | 1.000 | 1.000 | .993 | .865 | .632 | .393 | .181 | .095 | .049 | .020 | .010 |
| 4 | 1.000 | 1.000 | 1.000 | .982 | .865 | .632 | .330 | .181 | .095 | .039 | .020 |
| 10 | 1.000 | 1.000 | 1.000 | 1.000 | .993 | .918 | .632 | .393 | .221 | .095 | .049 |
| 20 | 1.000 | 1.000 | 1.000 | 1.000 | 1.000 | .993 | .865 | .632 | .393 | .181 | .095 |
| 40 | 1.000 | 1.000 | 1.000 | 1.000 | 1.000 | 1.000 | .982 | .865 | .632 | .330 | .181 |
| 100 | 1.000 | 1.000 | 1.000 | 1.000 | 1.000 | 1.000 | 1.000 | .993 | .918 | .632 | .393 |
| 200 | 1.000 | 1.000 | 1.000 | 1.000 | 1.000 | 1.000 | 1.000 | 1.000 | .993 | .865 | .632 |

Note that, in this table, both $\Delta t$ and T must be measured in the same time units. Hall and Howell (1) have extended this type of table to time intervals of 5 to 35 days. Since the probabilities in such tables are cumulative, by taking differences it is possible to compute probabilities of one or more occurrences of events with return periods between different values. For example the probability of one or more occurrences within 10 days of flood events with return periods between 10 and 100 years is 0.0024.

Under the assumptions that flood exceedences are independent identically distributed random variables and that the counting process for exceedence is a nonhomogeneous Poisson process, Todorovic and Woolhiser (2) have derived a one-dimensional distribution function for the time of occurrence of the largest event in some time interval.

If z(t) represents the number of flood peak exceedences in the time interval (0,t) and $\Lambda(t)$ is the expected value of z(t)

$$\Lambda(t) = E\left\{z(t)\right\} \tag{4-13}$$

then the probability that the time of occurrence T(t) of the largest momentary flood exceedence in the time interval (0,t) will be less than or equal to U, $P(T(t) \leq U)$, is given by

$$P(T(t) \leq U) = \exp\left\{-\Lambda(t)\right\} + \frac{\Lambda(U)}{\Lambda(t)}\left(1 - \exp\left\{-\Lambda(t)\right\}\right) \tag{4-14}$$

The expression $\Lambda(t)$ was derived by Todorovic and Woolhiser as a finite Fourier series.

Risk is discussed in more detail in a later chapter but it may be pointed out in passing that since from Equation 4-6, using $\lambda = \Delta t/T$,

$$p(o) = e^{-\Delta t/T} \tag{4-15}$$

then the return period, T, of a design flood with a risk of failure p(0) in a project life $\Delta t$ is:

$$T = \Delta t/\ln p(o) \tag{4-16}$$

e.g. for a 5% risk of failure (i.e. risk of an event of given magnitude occurring) in a 50 year life the project must be designed for a flood with return period of 975 years.

## References

1. Hall, W. A., and D. T. Howell, 1963, Estimating Flood Probabilities Within Specific Time Intervals, J. Hydrol., Vol. 1, No. 1, pp. 265-271.

2. Todorovic, P., and D. A. Woolhiser, 1972, On the Time when the Extreme Flood Occurs, Wat. Res. Res., Vol. 8, No. 6, pp. 1433-1438.

3. Weatherburn, C. E., 1962, A First Course in Mathematical Statistics, Cambridge University Press.

# CHAPTER 5

## NORMAL DISTRIBUTION

### Introduction

A distribution is said to be normal if the variable can take any value from - ∞ to + ∞ and the probability density function is defined as:

$$p(x) = \frac{1}{\sigma\sqrt{2\pi}} e^{\frac{-(x-\mu)^2}{2\sigma^2}} \qquad (5\text{-}1)$$

where $\mu$ and $\sigma$ are the distribution parameters shown later to be the population mean and standard deviation of the variable. The normal distribution is applicable if:

1. The variable is continuous.
2. Consecutive values are independent.
3. Probabilities are stable.

The normal distribution can be shown (5) to be a limiting case of the binomial when $p \to q \to \frac{1}{2}$ and $n \to \infty$. One of the features of the normal distribution is that the mean, mode and median are all the same.

If the variable, x, is standardized, i.e. forced to a mean of zero and unit variance by subtracting the mean and dividing by the standard deviation, and is denoted by t then Equation 5-1 becomes

$$p(t) = \frac{1}{\sqrt{2\pi}} e^{-t^2/2} \qquad (5\text{-}2)$$

which is known as the standard normal distribution. Equation 5-2 has been approxomated (accuracy > $2.27 \times 10^{-3}$) by a series of polynomials (1) such as:

$$p(t) = (a_o + a_1 t^2 + a_2 t^4 + a_3 t^6)^{-1} \qquad (5\text{-}3)$$

where

$$a_o = 2.490895, \quad a_2 = -0.024393,$$
$$a_1 = 1.466003, \quad a_3 = 0.178257$$

Tables of the ordinates of the normal curve are also available such as Table 5-1.

The probability corresponding to any interval in the range of the variate is represented by the area under the probability

Table 5-1

Ordinates of the Normal Curve

| t | 0.00 | 0.01 | 0.02 | 0.03 | 0.04 | 0.05 | 0.06 | 0.07 | 0.08 | 0.09 |
|---|---|---|---|---|---|---|---|---|---|---|
| 0.0 | .3989 | .3989 | .3989 | .3988 | .3986 | .3984 | .3982 | .3980 | .3977 | .3973 |
| 0.1 | .3970 | .3965 | .3961 | .3956 | .3951 | .3945 | .3939 | .3932 | .3925 | .3918 |
| 0.2 | .3910 | .3902 | .3894 | .3885 | .3876 | .3867 | .3857 | .3847 | .3836 | .3825 |
| 0.3 | .3814 | .3802 | .3790 | .3778 | .3765 | .3752 | .3739 | .3725 | .3712 | .3697 |
| 0.4 | .3683 | .3668 | .3653 | .3637 | .3621 | .3605 | .3589 | .3572 | .3555 | .3538 |
| 0.5 | .3521 | .3503 | .3485 | .3467 | .3448 | .3429 | .3410 | .3391 | .3372 | .3352 |
| 0.6 | .3332 | .3312 | .3292 | .3271 | .3251 | .3230 | .3209 | .3187 | .3166 | .3144 |
| 0.7 | .3123 | .3101 | .3079 | .3056 | .3034 | .3011 | .2989 | .2966 | .2943 | .2920 |
| 0.8 | .2897 | .2874 | .2850 | .2827 | .2803 | .2780 | .2756 | .2732 | .2709 | .2685 |
| 0.9 | .2661 | .2637 | .2613 | .2589 | .2565 | .2541 | .2516 | .2492 | .2468 | .2444 |
| 1.0 | .2420 | .2396 | .2371 | .2347 | .2323 | .2299 | .2275 | .2251 | .2227 | .2203 |
| 1.1 | .2179 | .2155 | .2131 | .2107 | .2083 | .2059 | .2036 | .2012 | .1989 | .1965 |
| 1.2 | .1942 | .1919 | .1895 | .1872 | .1849 | .1826 | .1804 | .1781 | .1758 | .1736 |
| 1.3 | .1714 | .1691 | .1669 | .1647 | .1626 | .1604 | .1582 | .1561 | .1539 | .1518 |
| 1.4 | .1497 | .1476 | .1456 | .1435 | .1415 | .1394 | .1374 | .1354 | .1334 | .1315 |
| 1.5 | .1295 | .1276 | .1257 | .1238 | .1219 | .1200 | .1182 | .1163 | .1145 | .1127 |
| 1.6 | .1109 | .1092 | .1074 | .1057 | .1040 | .1023 | .1006 | .0989 | .0973 | .0957 |
| 1.7 | .0940 | .0925 | .0909 | .0893 | .0878 | .0863 | .0848 | .0833 | .0818 | .0804 |
| 1.8 | .0790 | .0775 | .0761 | .0748 | .0734 | .0721 | .0707 | .0694 | .0681 | .0669 |
| 1.9 | .0656 | .0644 | .0632 | .0620 | .0608 | .0596 | .0584 | .0573 | .0562 | .0551 |
| 2.0 | .0540 | .0529 | .0519 | .0508 | .0498 | .0488 | .0478 | .0468 | .0459 | .0449 |
| 2.1 | .0440 | .0431 | .0422 | .0413 | .0404 | .0395 | .0387 | .0379 | .0371 | .0363 |
| 2.2 | .0355 | .0347 | .0339 | .0332 | .0325 | .0317 | .0310 | .0303 | .0297 | .0290 |
| 2.3 | .0283 | .0277 | .0270 | .0264 | .0258 | .0252 | .0246 | .0241 | .0235 | .0229 |
| 2.4 | .0224 | .0219 | .0213 | .0208 | .0203 | .0198 | .0194 | .0189 | .0184 | .0180 |
| 2.5 | .0175 | .0171 | .0167 | .0163 | .0158 | .0154 | .0151 | .0147 | .0143 | .0139 |
| 2.6 | .0136 | .0132 | .0129 | .0126 | .0122 | .0119 | .0116 | .0113 | .0110 | .0107 |
| 2.7 | .0104 | .0101 | .0099 | .0096 | .0093 | .0091 | .0088 | .0086 | .0084 | .0081 |
| 2.8 | .0079 | .0077 | .0075 | .0073 | .0071 | .0069 | .0067 | .0065 | .0063 | .0061 |
| 2.9 | .0060 | .0058 | .0056 | .0055 | .0053 | .0051 | .0050 | .0048 | .0047 | .0046 |
| 3.0 | .0044 | .0043 | .0042 | .0040 | .0039 | .0038 | .0037 | .0036 | .0035 | .0034 |
| 3.1 | .0033 | .0032 | .0031 | .0030 | .0029 | .0028 | .0027 | .0026 | .0025 | .0025 |
| 3.2 | .0024 | .0023 | .0022 | .0022 | .0021 | .0020 | .0020 | .0019 | .0018 | .0018 |
| 3.3 | .0017 | .0017 | .0016 | .0016 | .0015 | .0015 | .0014 | .0014 | .0013 | .0013 |
| 3.4 | .0012 | .0012 | .0012 | .0011 | .0011 | .0010 | .0010 | .0010 | .0009 | .0009 |
| 3.5 | .0009 | .0008 | .0008 | .0008 | .0008 | .0007 | .0007 | .0007 | .0007 | .0006 |
| 3.6 | .0006 | .0006 | .0006 | .0005 | .0005 | .0005 | .0005 | .0005 | .0005 | .0004 |
| 3.7 | .0004 | .0004 | .0004 | .0004 | .0004 | .0004 | .0003 | .0003 | .0003 | .0003 |
| 3.8 | .0003 | .0003 | .0003 | .0003 | .0003 | .0002 | .0002 | .0002 | .0002 | .0002 |
| 3.9 | .0002 | .0002 | .0002 | .0002 | .0002 | .0002 | .0002 | .0002 | .0001 | .0001 |

Note: $t = (x-\mu)/\sigma$

density curve,

$$P(x) = \int_{-\infty}^{x} \frac{1}{\sigma\sqrt{2\pi}} e^{\frac{-(x-\mu)^2}{2\sigma^2}} dx \qquad (5\text{-}4)$$

Standardizing, the cumulative probability corresponding to Equation 5-2 is:

$$P(t) = \int_{-\infty}^{t} \frac{1}{\sqrt{2\pi}} e^{-t^2/2} \, dt \qquad (5\text{-}5)$$

Similarly, Abramowitz and Stegun (1) list several approximations for Equations 5-5. A convenient polynomial approximation with an error term less than $1 \times 10^{-5}$ is

$$P(t) = 1 - f(t)(a_1 q + a_2 q^2 + a_3 q^3) \qquad (5\text{-}6)$$

where q is 1.0/(1.0 + a t), t is the positive standard normal deviate and a, $a_1$, $a_2$ and $a_3$ are constants with values

$$a = 0.33267 \qquad a_2 = -0.12017$$

$$a_1 = 0.43618 \qquad a_3 = 0.93730$$

A similar polynomial approximation has been programmed for computer by IBM (3) as subroutine NDTR. Tables of the area under the standard normal curve are also available such as Table 5-2.

## Estimation of Parameters

The estimations of parameters for the normal distribution by the methods of moments and maximum likelihood have been used as examples of techniques in Chapter 3. The results may be summarized as

$$\mu = \mu_1^1 \qquad (5\text{-}7)$$

$$\sigma = \sqrt{\mu_2} \qquad (5\text{-}8)$$

All odd central moments are zero and all even central moments may be expressed in terms of $\mu_2$ as

$$\mu_{2r} = \frac{(2r)!}{2^r r!} \mu_2^r \qquad (5\text{-}9)$$

e.g.

$$\mu_4 = 3\mu_2^2 = 3\sigma^4 \qquad (5\text{-}10)$$

## Frequency Factor

In the case of the normal distribution the frequency factor, K, in the standard equation

$$x_T = \mu + K\sigma \qquad (5\text{-}11)$$

is the standard normal deviate, t.

Table 5-2

Area Under the Standard Normal Curve

| t | 0.00 | 0.01 | 0.02 | 0.03 | 0.04 | 0.05 | 0.06 | 0.07 | 0.08 | 0.09 |
|---|---|---|---|---|---|---|---|---|---|---|
| 0.0 | .0000 | .0040 | .0080 | .0120 | .0159 | .0199 | .0239 | .0279 | .0319 | .0359 |
| 0.1 | .0398 | .0438 | .0478 | .0517 | .0557 | .0596 | .0636 | .0675 | .0714 | .0753 |
| 0.2 | .0793 | .0832 | .0871 | .0910 | .0948 | .0987 | .1026 | .1064 | .1103 | .1141 |
| 0.3 | .1179 | .1217 | .1255 | .1293 | .1331 | .1368 | .1406 | .1443 | .1480 | .1517 |
| 0.4 | .1554 | .1591 | .1628 | .1664 | .1700 | .1736 | .1772 | .1808 | .1844 | .1879 |
| 0.5 | .1915 | .1950 | .1985 | .2019 | .2054 | .2088 | .2123 | .2157 | .2190 | .2224 |
| 0.6 | .2257 | .2291 | .2324 | .2357 | .2389 | .2422 | .2454 | .2486 | .2518 | .2549 |
| 0.7 | .2580 | .2611 | .2642 | .2673 | .2704 | .2734 | .2764 | .2794 | .2823 | .2852 |
| 0.8 | .2881 | .2910 | .2939 | .2967 | .2995 | .3023 | .3051 | .3078 | .3106 | .3133 |
| 0.9 | .3159 | .3186 | .3212 | .3238 | .3264 | .3289 | .3315 | .3340 | .3365 | .3389 |
| 1.0 | .3413 | .3438 | .3461 | .3485 | .3508 | .3531 | .3554 | .3577 | .3599 | .3621 |
| 1.1 | .3643 | .3665 | .3686 | .3708 | .3729 | .3749 | .3770 | .3790 | .3810 | .3830 |
| 1.2 | .3849 | .3869 | .3888 | .3907 | .3925 | .3944 | .3962 | .3980 | .3997 | .4015 |
| 1.3 | .4032 | .4049 | .4066 | .4082 | .4099 | .4115 | .4131 | .4147 | .4162 | .4177 |
| 1.4 | .4192 | .4207 | .4222 | .4236 | .4251 | .4265 | .4279 | .4292 | .4306 | .4319 |
| 1.5 | .4332 | .4345 | .4357 | .4370 | .4382 | .4394 | .4406 | .4418 | .4430 | .4441 |
| 1.6 | .4452 | .4463 | .4474 | .4485 | .4495 | .4505 | .4515 | .4525 | .4535 | .4545 |
| 1.7 | .4554 | .4564 | .4573 | .4582 | .4591 | .4599 | .4608 | .4616 | .4625 | .4633 |
| 1.8 | .4641 | .4649 | .4656 | .4664 | .4671 | .4678 | .4686 | .4693 | .4699 | .4706 |
| 1.9 | .4713 | .4719 | .4726 | .4732 | .4738 | .4744 | .4750 | .4756 | .4762 | .4767 |
| 2.0 | .4772 | .4778 | .4783 | .4788 | .4793 | .4798 | .4803 | .4808 | .4812 | .4817 |
| 2.1 | .4821 | .4826 | .4830 | .4835 | .4838 | .4842 | .4846 | .4850 | .4854 | .4857 |
| 2.2 | .4861 | .4865 | .4868 | .4871 | .4875 | .4878 | .4881 | .4884 | .4887 | .4890 |
| 2.3 | .4893 | .4896 | .4898 | .4901 | .4904 | .4906 | .4909 | .4911 | .4913 | .4916 |
| 2.4 | .4918 | .4920 | .4922 | .4925 | .4927 | .4929 | .4931 | .4932 | .4934 | .4936 |
| 2.5 | .4938 | .4940 | .4941 | .4943 | .4945 | .4946 | .4948 | .4949 | .4951 | .4952 |
| 2.6 | .4953 | .4955 | .4956 | .4957 | .4959 | .4960 | .4961 | .4962 | .4963 | .4964 |
| 2.7 | .4965 | .4966 | .4967 | .4968 | .4969 | .4970 | .4971 | .4972 | .4973 | .4974 |
| 2.8 | .4974 | .4975 | .4976 | .4977 | .4977 | .4978 | .4979 | .4980 | .4980 | .4981 |
| 2.9 | .4981 | .4982 | .4983 | .4983 | .4984 | .4984 | .4985 | .4985 | .4986 | .4986 |
| 3.0 | .4986 | .4987 | .4987 | .4988 | .4988 | .4989 | .4989 | .4989 | .4990 | .4990 |
| 3.1 | .4990 | .4991 | .4991 | .4991 | .4992 | .4992 | .4992 | .4992 | .4993 | .4993 |

Note: $t = (x-\mu)/\sigma$

The easiest way of obtaining the standard normal deviate is from tables of the area under the normal curve such as Table 5-2. As an example consider the determination of t for an event with a mean return period of 100 years. The cumulative probability of non-exceedence, as a percentage, associated with this T value is 99%, (1-1/T). Of this, 50% is contributed by the integral of the standard normal density curve from $-\infty$ to 0 leaving 49% from the integral from 0 to x. Looking up 0.49 in the body of the table it is found that t is given as 2.33. The corresponding event magnitude is therefore $\mu + 2.33\sigma$.

This is using the standard normal curve in a one-tail manner. Statistical tests often use the standard normal deviate corresponding to a two-tail situation; that is, one in which the event might occur at either tail of the distribution. Figures 5-1 and 5-2 compare these two concepts for an area under the curve of 0.95.

As a convenience, the frequency factors (standard normal deviates) for the normal (and lognormal) distribution are given in Table 5-3 for some commonly used cumulative probabilities. Table 5-3 also shows the standard normal deviates for the two-tail situation.

Table 5-3

Frequency Factor for Use in Normal and Lognormal Distributions

| Cumulative Probability, P, % | | | | | |
|---|---|---|---|---|---|
| 50 | 80 | 90 | 95 | 98 | 99 |
| **Corresponding Return, Period, T, Years** | | | | | |
| 2 | 5 | 10 | 20 | 50 | 100 |
| **Frequency Factor** | | | | | |
| 0 | 0.8416 | 1.2816 | 1.6449 | 2.0538 | 2.3264 |
| **Two-Tail Standard Normal Deviate** | | | | | |
| 0.6745 | 1.2816 | 1.6449 | 1.9600 | 2.3264 | 2.5758 |

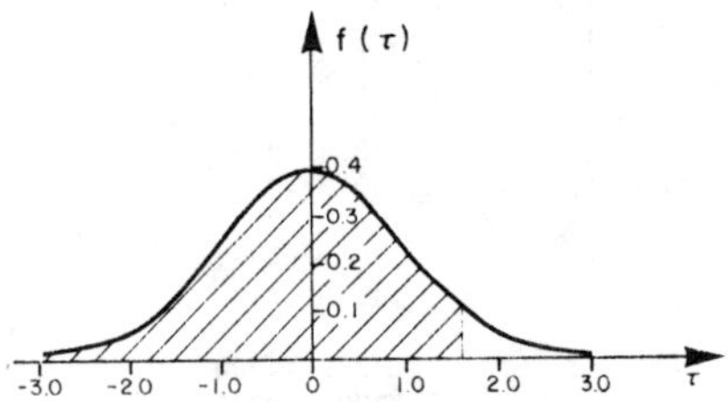


Figure 5-1. Area under the standard normal curve, one-tail.

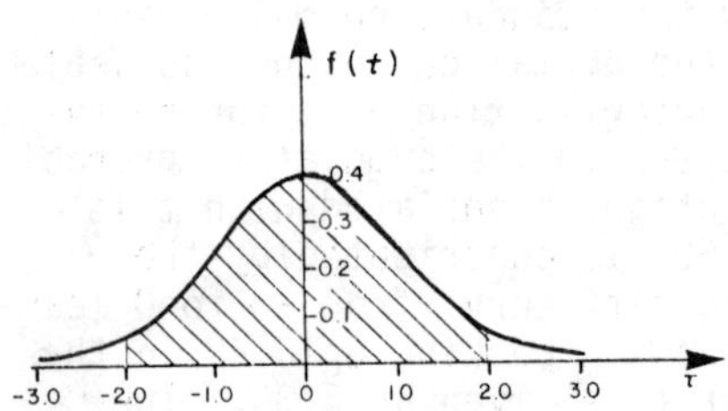


Figure 5-2. Area under the standard normal curve, two-tail.

It is sometimes useful to combine tables of the standard normal deviate with the plotting position m/(n+1) in which m is the rank of the event in decreasing order of magnitude. Table 5-4 is a sample table of this type.

Polynomial approximations are also available (1) to obtain the standard normal deviate corresponding to a given probability is P(t), then for $0 < P(t) \leq 0.5$

$$t \simeq w - \frac{c_0+c_1w+c_2w^2}{1+d_1w+d_2w^2+d_3w^3} \tag{5-12}$$

where

$c_0 = 2.515517$; $d_1 = 1.432788$

$c_1 = 0.802853$; $d_2 = 0.189269$

$c_2 = 0.010328$; $d_3 = 0.001308$

Table 5-4

Plotting Positions, P, and Standard Normal Deviate, t, for a Range of Samples of Size n Events

| Event Rank No. (m) | n = 19 P % | t | n = 20 P % | t | n = 21 P % | t | n = 22 P % | t | n = 23 P % | t |
|---|---|---|---|---|---|---|---|---|---|---|
| 1 | 5.0 | 1.64 | 4.8 | 1.67 | 4.5 | 1.69 | 4.3 | 1.71 | 4.3 | 1.73 |
| 2 | 10.0 | 1.28 | 9.5 | 1.31 | 9.1 | 1.34 | 8.7 | 1.36 | 8.3 | 1.38 |
| 3 | 15.0 | 1.04 | 14.3 | 1.07 | 13.6 | 1.10 | 13.0 | 1.12 | 12.5 | 1.15 |
| 4 | 20.0 | .84 | 19.0 | .87 | 18.2 | .91 | 17.4 | .94 | 16.7 | .97 |
| 5 | 25.0 | .67 | 23.8 | .71 | 22.7 | .75 | 21.7 | .78 | 20.8 | .81 |
| 6 | 30.0 | .52 | 28.6 | .57 | 27.3 | .61 | 26.1 | .64 | 25.0 | .67 |
| 7 | 35.0 | .39 | 33.3 | .43 | 31.8 | .47 | 30.4 | .51 | 29.2 | .55 |
| 8 | 40.0 | .25 | 38.1 | .30 | 36.4 | .35 | 34.8 | .39 | 33.3 | .43 |
| 9 | 45.0 | .13 | 42.9 | .18 | 40.9 | .23 | 39.1 | .27 | 37.5 | .32 |
| 10 | 50.0 | .00 | 47.6 | .06 | 45.5 | .12 | 43.5 | .16 | 41.7 | .21 |
| 11 | 55.0 | - .13 | 52.4 | - .06 | 50.0 | .00 | 47.8 | .05 | 45.8 | .10 |
| 12 | 60.0 | - .25 | 57.1 | - .18 | 54.5 | - .12 | 52.2 | - .05 | 50.0 | .00 |
| 13 | 65.0 | - .39 | 61.9 | - .30 | 59.1 | - .23 | 56.5 | - .16 | 54.2 | - .10 |
| 14 | 70.0 | - .52 | 66.7 | - .43 | 63.6 | - .35 | 60.9 | - .27 | 58.3 | - .21 |
| 15 | 75.0 | - .67 | 71.4 | - .57 | 68.2 | - .47 | 65.2 | - .39 | 62.5 | - .32 |
| 16 | 80.0 | - .84 | 76.2 | - .71 | 72.7 | - .61 | 69.6 | - .51 | 66.7 | - .43 |
| 17 | 85.0 | -1.04 | 81.0 | - .87 | 77.3 | - .75 | 73.9 | - .64 | 70.8 | - .55 |
| 18 | 90.0 | -1.28 | 85.7 | -1.07 | 81.8 | - .91 | 78.3 | - .78 | 75.0 | - .67 |
| 19 | 95.0 | -1.64 | 90.5 | -1.31 | 86.4 | -1.10 | 82.6 | - .94 | 79.2 | - .81 |
| 20 | | | 95.2 | -1.67 | 90.9 | -1.34 | 87.0 | -1.12 | 83.3 | - .97 |
| 21 | | | | | 95.5 | -1.69 | 91.3 | -1.36 | 87.5 | -1.15 |
| 22 | | | | | | | 95.7 | -1.71 | 91.7 | -1.38 |
| 23 | | | | | | | | | 95.8 | -1.73 |

Notes: $P = m/(n+1)$, $t = (x-\mu)/\sigma$

and where

$$w = \sqrt{\ln (1/P(t)^2)} \qquad (5\text{-}13)$$

This approximation is particularly useful when digital computers are used since it avoids the reverse integration in Equation 5-5. The error term in the approximation is stated (1) to be less than $4.5 \times 10^{-4}$. For values of $P(t) > 0.5$ use $1.0 - P(t)$ in the approximation and change the sign of the calculated value of t. This approximation has been programmed for computer as subroutine NDTRI in the IBM Scientific Subroutine Package (3).

## Standard Error

*Method of Moments* - The general equation for the parameter $\delta$ in the standard error of estimate of a 2 parameter distribution from the method of moments (see Equation 3-51) is:

$$\delta = \left\{1 + \gamma_1 K + \left[\gamma_2 - 1\right] \frac{K^2}{4}\right\}^{\frac{1}{2}} \tag{5-14}$$

For the normal distribution the central moments are given by:

$$\mu_2 = \sigma^2 \tag{5-15}$$

$$\mu_3 = 0 \tag{5-16}$$

and

$$\mu_4 = 3\sigma^4 \tag{5-17}$$

Substituting these moment expressions into 5-14 and using t as frequency factor results in

$$\delta = [1 + t^2/2]^{\frac{1}{2}} \tag{5-18}$$

so that the standard error is

$$S_T = \delta \sqrt{\frac{\sigma^2}{n}} \tag{5-19}$$

Table 5-5 provides values of the parameter $\delta$ for some common cumulative probabilities.

Table 5-5

Parameter $\delta$ for Use in Standard Error of Normal and Lognormal Distributions

| Cumulative Probability, P, % | | | | | |
|---|---|---|---|---|---|
| 50 | 80 | 90 | 95 | 98 | 99 |
| **Corresponding Return Period, T, Years** | | | | | |
| 2 | 5 | 10 | 20 | 50 | 100 |
| 1.0000 | 1.1638 | 1.3497 | 1.5340 | 1.7634 | 1.9249 |

An alternate method of tabulating standard error is as the ratio $S_T/\sigma$. Hardison (2) has provided tables of this type from which Table 5-6 has been derived. Since both $\delta$ and the ratio of deviations $S_T/\sigma$ are dimensionless, Tables 5-5 and 5-6 can equally well be used with the lognormal distribution.

If v(K) is the expected coefficient of variation of the estimate $x_T$ and V(K) is the coefficient of variation of the true event $y_T$ defined respectively as

$$v(K) = S_T/y_T \tag{5-20}$$

Table 5-6

Dimensionless Ratio of the Standard Error of the T-Year Event to the Standard Deviation of the Annual Events for Normal and Lognormal Distributions

| Return Period T | Sample Length, n | | | | | |
|---|---|---|---|---|---|---|
| | 2 | 5 | 10 | 20 | 50 | 100 |
| 2 | 0.707 | 0.447 | 0.316 | 0.224 | 0.141 | 0.100 |
| 5 | 0.782 | 0.495 | 0.350 | 0.247 | 0.156 | 0.116 |
| 10 | 0.954 | 0.604 | 0.427 | 0.302 | 0.191 | 0.135 |
| 20 | 1.083 | 0.685 | 0.484 | 0.342 | 0.217 | 0.153 |
| 50 | 1.208 | 0.764 | 0.540 | 0.382 | 0.242 | 0.176 |
| 100 | 1.364 | 0.863 | 0.610 | 0.431 | 0.273 | 0.193 |

and

$$V(K) = \sigma/\mu \tag{5-21}$$

then from Equation 5-19

$$v(K) = \frac{V(K)}{\sqrt{n}} \sqrt{\frac{1 + t^2/2}{(1 + t.V(K))}} \tag{5-22}$$

Nash and Amorocho (4) have shown that as $K \to \infty$, $v(K) \to 1/\sqrt{2n}$. Graphs of $v(K)\sqrt{2n}$ versus t for different values of V(K) show that the coefficient of variation has a minimum value between t = 0 and t = 2 and that for values of V(K) > 0.2 the mean annual event (t = 0) is less well defined than the event corresponding to t = 1.

*Maximum Likelihood* - For the normal distribution with two parameters (using $\alpha$ and $\beta^2$ for generality) the T-year event is a function of these parameters and t:

$$x_T = f(\alpha, \beta^2, t) \tag{5-23}$$

So that the variance of $x_T$, $S_T^2$, is

$$S_T^2 = \left(\frac{\partial x}{\partial \alpha}\right)^2 \text{var } \alpha + \left(\frac{\partial x}{\partial \beta^2}\right)^2 \text{var } \beta^2 + 2\frac{\partial x}{\partial \alpha}\frac{\partial x}{\partial \beta} \text{cov } (\alpha, \beta^2) \tag{5-24}$$

The result is obvious but for completeness we follow the procedure through.

From the logarithm of the likelihood equation (Equation 3-17) the required partial derivatives are

$$\frac{\partial^2 \ln L}{\partial \alpha^2} = -\frac{n}{\beta^2} \tag{5-25}$$

$$\frac{\partial^2 \ln L}{\partial(\beta^2)^2} = \frac{n}{2\beta^4} - \frac{\sum_{i=1}^{n}(x_i - \alpha)^2}{2\beta^6} \quad (5\text{-}26)$$

$$\frac{\partial^2 \ln L}{\partial\alpha\partial\beta} = -\frac{\sum_{n=1}^{n}(x_i - \alpha)}{\beta^4} \quad (5\text{-}27)$$

but from the maximum likelihood solutions

$$\sum_{i=1}^{n}(x_i - \alpha) = 0 \quad (5\text{-}28)$$

and

$$\sum_{i=1}^{n}(x_i - \alpha)^2 = n\beta^2 \quad (5\text{-}29)$$

so that

$$\frac{\partial^2 \ln L}{\partial(\beta^2)^2} = -\frac{n}{2\beta^4} \quad (5\text{-}30)$$

and

$$\frac{\partial^2 \ln L}{\partial\alpha\partial\beta} = 0 \quad (5\text{-}31)$$

Following the procedure of taking minors and determinants of the likelihood matrix the variances and covariences of the parameters are

$$\text{var}\ \alpha = \beta^2/n \quad (5\text{-}32)$$

$$\text{var}\ \beta^2 = 2\beta^4/n \quad (5\text{-}33)$$

and

$$\text{cov}\ (\alpha, \beta^2) = 0 \quad (5\text{-}34)$$

Now the partial derivatives in Equation 5-24 can be obtained from the frequency equation for the normal distribution as:

$$\frac{\partial x}{\partial\alpha} = 1 \quad (5\text{-}35)$$

$$\frac{\partial x}{\partial\beta^2} = t/2\beta \quad (5\text{-}36)$$

The standard error of the T-year event is then:

$$S_T^2 = \frac{\beta^2}{n}\left[1 + t^2/2\right] \quad (5\text{-}37)$$

which is the same result as by method of moments.

As an example of the computation of a confidence interval for a normal distribution consider the estimate of the 100 year event from a sample record of 50 years. The cumulative probability of non-exceedence of the 100 year event (area under the normal curve) is 99% and so from Table 5-2 the standard normal deviate, t, is 2.33. In Equation 5-19, $S_T$ is computed as 0.273$\sigma$, where $\sigma$ is the sample estimate of the population standard deviation. Using a 95% confidence level (two-tail) the confidence interval around the 100 year event, $x_{100}$, is given as $x_{100} \pm 1.96 * 0.273\sigma$.

## References


1. Abramowitz, M., and I. A. Stegun, 1965, Handbook of Mathematical Functions, Dover Publications, New York.

2. Hardison, C. H., 1969, Accuracy of Streamflow Characteristics, USGS Professional Paper No. 650-D, pp. D210-D214.

3. IBM, 1968, System/360 Scientific Subroutine Package (360A-CM-03X) Version III, Programmer's Manual, White Plains, New York.

4. Nash, J. E., and J. Amorocho, 1966, The Accuracy of the Prediction of Floods of High Return Period, Wat. Res. Res., Vol. 2, No. 2, pp. 191-198.

5. Weatherburn, C. E., 1962, A First Course in Mathematical Statistics, Chambridge University Press, Cambridge, England.

## CHAPTER 6

## TWO PARAMETER LOGNORMAL DISTRIBUTION

### Introduction

If the logarithms, ln x, of a variable x are normally distributed, then the variable x is said to be logarithmic-normally distributed so that

$$p(x) = \frac{1}{x\sigma_y\sqrt{2\pi}} e^{-\frac{[\ln x - \mu_y]^2}{2\sigma_y^2}} \qquad (6\text{-}1)$$

where $\mu_y$ and $\sigma_y$ are the mean and standard deviation of the natural logarithms of x.

Chow (2) has provided a theoretical justification for the use of the lognormal distribution. The causative factors for many hydrologic variables act multiplicatively rather than additively and so the logarithms of these factors will satisfy the four basic conditions for normal distributions. The hydrologic variable will then be the product of these causative factors.

Singh and Sinclair (9) described a mixed, or compound, probability distribution made up of two lognormal distributions, as:

$$P(x) = a_1P_1(x) + a_2P_2(x) \qquad (6\text{-}2)$$

where

$$P_1(x) = \frac{1}{\sigma_1\sqrt{2\pi}} \int_{-\infty}^{x} e^{-(x - \mu_1)^2/2\sigma_1^2} dx \qquad (6\text{-}3)$$

$$P_2(x) = \frac{1}{\sigma_2\sqrt{2\pi}} \int_{-\infty}^{x} e^{-(x - \mu_2)^2/2\sigma_2^2} dx \qquad (6\text{-}4)$$

and

$$a_1 + a_2 = 1 \qquad (6\text{-}5)$$

The mean, variance and coefficient of skew of the distribution P(x) may be estimated from the sample, or computed from

$$\mu = a_1\mu_1 + a_2\mu_2 \qquad (6\text{-}6)$$

$$\sigma^2 = a_1\sigma_1^2 + a_2\sigma_2^2 + a_1a_2(\mu_2 - \mu_1)^2 \qquad (6\text{-}7)$$

$$\gamma_1 = \frac{3a_1a_2(\mu_1 - \mu_2)(\sigma_1^2 - \sigma_2^2)}{\sigma^3} + \frac{a_1a_2(a_2 - a_1)(\mu_1 - \mu_2)^3}{\sigma^3} \quad (6\text{-}8)$$

The advantage of this method (9) is that it has the versatility of high parameter models without the errors and uncertainties which result from the use of higher order sample moments.

Estimation of Parameters

*Method of Moments* - Applying the general equation for moments about the origin to the pdf of the lognormal distribution:

$$\mu'_r = \int_0^\infty x^r \frac{1}{x\sigma_y \sqrt{2\pi}} \exp\left[-(\ln x - \mu_y)^2/2\sigma_y^2\right] dx \quad (6\text{-}9)$$

Complete the square in the exponent by substituting

$$\omega = (\ln x - r\sigma_y^2 - \mu_y)/\sigma_y \quad (6\text{-}10)$$

so that

$$\mu'_r = e^{r\mu_y + r^2\sigma_y^2/2} \cdot \frac{1}{\sqrt{2\pi}} \int_{-\infty}^{\infty} e^{-\omega^2/2} d\omega \quad (6\text{-}11)$$

and since the latter integral is unity

$$\mu'_r = e^{r\mu_y + r^2\sigma_y^2/2} \quad (6\text{-}12)$$

for which

$$\mu'_1 = \mu = e^{\mu_y + \sigma_y^2/2} \quad (6\text{-}13)$$

and

$$\mu_2 = \sigma^2 = (e^{\sigma_y^2} - 1) \cdot \mu_1'^2 \quad (6\text{-}14)$$

Similarly the third central moment, $\mu_3$, can be shown to be

$$\mu_3 = (e^{3\sigma_y^2} - 3e^{\sigma_y^2} + 2)\, e^{3\mu_y + 3\sigma_y^2/2} \quad (6\text{-}15)$$

The coefficient of variation of the recorded events $\mu_2^{\frac{1}{2}}/\mu'_1$, is then

$$z = (e^{\sigma_y^2} - 1)^{\frac{1}{2}} \quad (6\text{-}16)$$

and the coefficient of skew of the recorded events $\mu_e/\mu_2^{3/2}$, is

$$\gamma_1 = \frac{e^{3\sigma_y^2} - 3e^{\sigma_y^2} + 2}{(e^{\sigma_y^2} - 1)^{3/2}} \quad (6\text{-}17)$$

From Equations 6-16 and 6-17 the relationship between the coefficients of variation and skew is

$$\gamma_1 = 3z + z^3 \quad (6\text{-}18)$$

*Maximum Likelihood* - Following the maximum likelihood procedure the logarithm of the likelihood expression for Equation 6-1 is

$$\ln L = -\frac{n}{2}\ln 2\pi - \frac{n}{2}\ln \sigma_y^2 + \sum_{i=1}^{n} \ln (1/x_i) - \sum_{i=1}^{n} (\ln x_i - \mu_y)^2/2\sigma_y^2 \quad (6\text{-}19)$$

Differentiating Equation 6-19 with respect to $\mu_y$ and $\sigma_y^2$ and equating to zero yields the maximum likelihood estimates

$$\mu_y = \sum_{i=1}^{n} \ln x_i/n \quad (6\text{-}20)$$

$$\sigma_y^2 = \sum_{i=1}^{n} (\ln x_i - \mu_y)^2/n \quad (6\text{-}21)$$

In practice a problem frequently arises in applying either the method of maximum likelihood or the method of moments. Many streamflow records have the occasional zero flow and, when taking logarithms, this becomes $-\infty$ and cannot be processed. Several simple solutions to this problem have been proposed (7) such as

1. Add 1.0 to all data.
2. Add small positive value (such as 0.1, 0.01, 0.001, etc.) to all data.
3. Substitute 1.0 in place of all zero readings.
4. Substitute small positive value in place of all zero readings.
5. Ignore all zero observations.

All of these solutions affect the parameters of the distribution (1 and 2 affect the mean, 3, 4 and 5 affect both the mean and variance). Substitution of small positive values is to be avoided because of the large effect this has in a logarithmic scale.

An alternate solution is to consider the probability distribution as the sum of a probability mass at zero and a probability density distribution over the remainder of the range (10), so that:

$$P[0 \leq x \leq x_o] = P[x=0] + P[0 < x \leq x_o] \qquad (6\text{-}22)$$

In this way, if $p_o$ is the probability of occurrence of x = 0 in any one year and $p_x$ is the conditional probability of occurence of x in any one year given that x is not zero, then:

$$P[0 < x < x_o] = p_x(1 - p_o) \qquad (6\text{-}23)$$

Jennings and Benson (6) have described an operational procedure of applying this conditional probability. Given a data sample containing zero events (or events below any base flow $Q_b$) to which it is required to fit a log-probability distribution, first of all fit the probability distribution to those events above $Q_b$. Then multiply the resulting probabilities by the ratio of the number of events greater than zero (or > $Q_b$) to the total number of events in the sample. The result will be the required probabilities of exceedence.

Frequency Factor

If y = ln x is normally distributed, the general frequency equation (e.g. Equation 3-30) can be written in terms of logarithms as:

$$\ln x_T = y_T = \mu_y + t\sigma_y \qquad (6\text{-}24)$$

where, t is the standard normal deviate.

Alternately to avoid computation of the mean and standard deviation of logarithms, the general frequency equation, Equation 3-30, may be modified by substituting for μ and σ from Equations 6-13 and 6-14 and using $e^{y_T}$ in place of $x_T$:

$$e^{y_T} = e^{\mu_y + \sigma_y^2/2}\,[1 + K(e^{\sigma_y^2} - 1)^{\frac{1}{2}}] \qquad (6\text{-}25)$$

From Equation 6-25 the frequency factor for the lognormal distribution, K, is given by

$$K = \frac{e^{y_T - \mu_y - \sigma_y^2/2} - 1}{(e^{\sigma_y^2} - 1)^{\frac{1}{2}}} \qquad (6\text{-}26)$$

But, from Equation 6-24

$$y_T - \mu_y = t\sigma_y \qquad (6\text{-}27)$$

which, when substituted in Equation 6-26, yields:

$$K = \frac{e^{\sigma_y t - \sigma_y^2/2} - 1}{(e^{\sigma_y^2} - 1)^{\frac{1}{2}}} \qquad (6\text{-}28)$$

where, as before, t is the standard normal deviate. Now, by using Equation 6-16 relating to the coefficient of variation, z, of the observed events to the standard deviation of the logarithms, $\sigma_y$, all logarithmic parameters can be avoided:

$$K = \frac{e^{[\ln(1+z^2)]^{\frac{1}{2}}t - [\ln(1+z^2)]/2} - 1}{z} \qquad (6\text{-}29)$$

Table 6-1 shows values of K computed from Equation 6-29 for some commonly used return periods and various values of z, the coefficient of variation. Chow (1,2) gave more comprehensive tables

Table 6-1

Frequency Factor for Lognormal Distribution

| Coefficient of Variation z | Cumulative Probability, P, % / Corresponding Return Period, T, Years: 50 / 2 | 80 / 5 | 90 / 10 | 95 / 20 | 98 / 50 | 99 / 100 |
|---|---|---|---|---|---|---|
| 0.0500 | -0.0250 | 0.8334 | 1.2965 | 1.6863 | 2.1341 | 2.4370 |
| 0.1000 | -0.0496 | 0.8222 | 1.3078 | 1.7247 | 2.2130 | 2.5489 |
| 0.1500 | -0.0738 | 0.8085 | 1.3156 | 1.7598 | 2.2899 | 2.6607 |
| 0.2000 | -0.0971 | 0.7926 | 1.3200 | 1.7911 | 2.3640 | 2.7716 |
| 0.2500 | -0.1194 | 0.7746 | 1.3209 | 1.8183 | 2.4348 | 2.8805 |
| 0.3000 | -0.1406 | 0.7547 | 1.3183 | 1.8414 | 2.5016 | 2.9866 |
| 0.3500 | -0.1604 | 0.7333 | 1.3126 | 1.8602 | 2.5638 | 3.0890 |
| 0.4000 | -0.1788 | 0.7106 | 1.3037 | 1.8746 | 2.6212 | 3.1870 |
| 0.4500 | -0.1957 | 0.6870 | 1.2920 | 1.8848 | 2.6734 | 3.2799 |
| 0.5000 | -0.2111 | 0.6626 | 1.2778 | 1.8909 | 2.7202 | 3.3673 |
| 0.5500 | -0.2251 | 0.6379 | 1.2613 | 1.8931 | 2.7615 | 3.4488 |
| 0.6000 | -0.2375 | 0.6129 | 1.2428 | 1.8915 | 2.7974 | 3.5241 |
| 0.6500 | -0.2485 | 0.5879 | 1.2226 | 1.8866 | 2.8279 | 3.5930 |
| 0.7000 | -0.2582 | 0.5631 | 1.2011 | 1.8786 | 2.8532 | 3.6556 |
| 0.7500 | -0.2667 | 0.5387 | 1.1784 | 1.8677 | 2.8735 | 3.7118 |
| 0.8000 | -0.2739 | 0.5148 | 1.1548 | 1.8543 | 2.8891 | 3.7617 |
| 0.8500 | -0.2801 | 0.4914 | 1.1306 | 1.8388 | 2.9002 | 3.8056 |
| 0.9000 | -0.2852 | 0.4686 | 1.1060 | 1.8212 | 2.9071 | 3.8437 |
| 0.9500 | -0.2895 | 0.4466 | 1.0810 | 1.8021 | 2.9103 | 3.8762 |
| 1.0000 | -0.2929 | 0.4254 | 1.0560 | 1.7815 | 2.9098 | 3.9035 |

which, however, require somewhat awkward interpolation to use for values of z. It is worth noting that in Chow's tables the first line, for a zero coefficient of skew, is equivalent to the normal distribution. Since the relationship between the coefficients of variation and skew (Equation 6-18) does not hold for the normal distribution the frequency factors are independent of the coefficient of variation (8) at that particular coefficient of skew.

There are then two possible procedures for using a lognormal distribution to estimate T-year event magnitudes. The standard normal deviate, t, may be used with the mean and standard deviation of the logarithms of the recorded events, or the frequency factor, K, may be used with the mean and standard deviation of the recorded events.

As an example of these two procedures consider the following data from Collier (3). Annual maximum discharges of the Saint John River at Fort Kent, New Brunswick for the 37 years 1927 to 1963 yield a mean discharge of 81,000 cfs, a standard deviation of 22,800 cfs and a coefficient of variation of 0.28. The mean and standard deviation of the logarithms of these 37 events are 11.263 and 0.284.

Using the first method, the magnitude of the 100-year event (t = 2.326) is given by

$$x_{100} = e^{11.263 \times 2.326 \times 0.286} = 151{,}000 \text{ cfs} \qquad (6\text{-}30)$$

Using the second method the frequency factor, K, from Table 6-1 is 2.944 so that

$$x_{100} = 81{,}000 + 2.944 \times 22{,}800 = 148{,}000 \text{ cfs} \qquad (6\text{-}31)$$

The two methods thus produce comparable results and so, if the lognormal distribution is to be used, there is nothing to gain from the extra work involved in taking logarithms and computing $\mu_y$ and $\sigma_y$.

<u>Standard Error</u>

*Method of Moments* - The general equation for the parameter $\delta$ in the standard error of estimate of a 2-parameter distribution by the method of moments is:

$$\delta = \left\{1 + \gamma_1 K + \left[\gamma_2 - 1\right] \frac{K^2}{4}\right\}^{\frac{1}{2}} \qquad (6\text{-}32)$$

If the moment ratios, $\gamma_1$ and $\gamma_2$, are now taken for logarithms so that the frequency factor is t, the standard normal deviate, then:

$$\delta = [1 + t^2/2]^{\frac{1}{2}} \qquad (6\text{-}33)$$

and the standard error of estimate, $S_T$, in logarithmic units is:

$$S_T = \delta\sigma_y/\sqrt{n} \qquad (6\text{-}34)$$

Values of the parameter $\delta$ are given in Table 6-2 for various commonly used return periods. From the standard error in logarithmic

Table 6-2

Parameter δ for Use in Standard Error of Normal and Lognormal Distributions

| Cumulative Probability, P, % | | | | | |
|---|---|---|---|---|---|
| 50 | 80 | 90 | 95 | 98 | 99 |
| Corresponding Return Period, T, Years | | | | | |
| 2 | 5 | 10 | 20 | 50 | 100 |
| 1.0000 | 1.1638 | 1.3495 | 1.5339 | 1.7632 | 1.9251 |

units, $S_T$, the positive and negative standard errors can be derived as

$$PSE = x_T(e^{S_T} - 1) \tag{6-35}$$

and

$$NSE = -x_T(e^{-S_T} - 1) \tag{6-36}$$

Hardison (5) has published graphs showing the variation of $S_T$ with the standard deviation of the logarithms, $\sigma_y$, and has provided tables to convert $S_T$ from logarithmic units back to the units of the basic data.

Alternatively, to avoid the computation of the standard deviation of the logarithms, use can be made of the moment ratios of the original events

$$\gamma_1 = \frac{\mu_3}{\mu_2^{3/2}} = \frac{e^{3\sigma_y^2} - 3e^{\sigma_y^2} + 2}{(e^{\sigma_y^2} - 1)^{3/2}} \tag{6-37}$$

and

$$\gamma_2 = \frac{\mu_4}{\mu_2^2} = \left(e^{\sigma_y^2/2}\right)^8 + 2\left(e^{\sigma_y^2/2}\right)^6 + 3\left(e^{\sigma_y^2/2}\right)^4 - 3 \tag{6-38}$$

Substituting Equations 6-37 and 6-38 into Equation 6-32 and making use of the relationship between the coefficient of variation, z, and the standard deviation of the logarithms, $\sigma_y$, (Equation 6-16) results in:

$$\delta_y = [1 + (z^3 + 3z)K + (z^8 + 6z^6 + 15z^4 + 16z^2 + 2)K^2/4]^{\frac{1}{2}} \tag{6-39}$$

where K, the frequency factor, is given by 6-29 so that

$$S_T = \delta_y \sigma/\sqrt{n} \qquad (6\text{-}40)$$

where $S_T$ is now in linear units. Table 6-3 provides values of $\delta_y$ for some commonly used values of return period and coefficient of variation. Confidence limits on the event magnitude may be computed by this method as:

$$x_T \pm t\delta_y \sigma/\sqrt{n} \qquad (6\text{-}41)$$

assuming the normality of the distribution of $x_T$.

Table 6-3

Parameter $\delta_y$ for Use in Standard Error of Lognormal Distribution

| Coefficient of Variation | Cumulative Probability, P, % | | | | | |
|---|---|---|---|---|---|---|
| | 50 | 80 | 90 | 95 | 98 | 99 |
| | Corresponding Return Period, T, Years | | | | | |
| Z | 2 | 5 | 10 | 20 | 50 | 100 |
| .05 | .9983 | 1.2161 | 1.4323 | 1.6442 | 1.9086 | 2.0967 |
| .10 | .9932 | 1.2698 | 1.5222 | 1.7682 | 2.0765 | 2.2977 |
| .15 | .9848 | 1.3240 | 1.6187 | 1.9054 | 2.2674 | 2.5296 |
| .20 | .9733 | 1.3782 | 1.7210 | 2.0556 | 2.4817 | 2.7937 |
| .25 | .9589 | 1.4322 | 1.8288 | 2.2184 | 2.7199 | 3.0914 |
| .30 | .9420 | 1.4853 | 1.9416 | 2.3936 | 2.9826 | 3.4242 |
| .35 | .9229 | 1.5376 | 2.0590 | 2.5811 | 3.2704 | 3.7937 |
| .40 | .9021 | 1.5887 | 2.1810 | 2.7811 | 3.5841 | 4.2017 |
| .45 | .8801 | 1.6387 | 2.3073 | 2.9935 | 3.9246 | 4.6501 |
| .50 | .8575 | 1.6873 | 2.4381 | 3.2187 | 4.2929 | 5.1409 |
| .55 | .8351 | 1.7348 | 2.5733 | 3.4570 | 4.6903 | 5.6764 |
| .60 | .8138 | 1.7811 | 2.7133 | 3.7090 | 5.1182 | 6.2592 |
| .65 | .7945 | 1.8263 | 2.8582 | 3.9753 | 5.5781 | 6.8920 |
| .70 | .7784 | 1.8705 | 3.0083 | 4.2566 | 6.0719 | 7.5778 |
| .75 | .7669 | 1.9139 | 3.1642 | 4.5538 | 6.6013 | 8.3199 |
| .80 | .7615 | 1.9566 | 3.3261 | 4.8678 | 7.1685 | 9.1217 |
| .85 | .7635 | 1.9986 | 3.4946 | 5.1996 | 7.7758 | 9.9870 |
| .90 | .7746 | 2.0402 | 3.6702 | 5.5503 | 8.4255 | 10.9198 |
| .95 | .7959 | 2.0815 | 3.8533 | 5.9210 | 9.1202 | 11.9242 |
| 1.00 | .8284 | 2.1225 | 4.0445 | 6.3129 | 9.8625 | 13.0046 |

*Maximum Likelihood* - The general expression for the maximum likelihood estimate of the standard error of the T-year event for the 2 parameter lognormal distribution is

$$S_T^2 = \left(\frac{\partial x_T}{\partial \mu_y}\right)^2 \text{var } \mu_y + \left(\frac{\partial x_T}{\partial \sigma_y^2}\right) \text{var } \sigma_y^2 + 2\frac{\partial x_T}{\partial \mu_y}\frac{\partial x_T}{\partial \sigma_y^2} \text{cov } (\mu_y, \sigma_y^2) \qquad (6\text{-}42)$$

and from Equation 6-24

$$x_T = e^{\mu_y + t\sigma_y} \qquad (6\text{-}43)$$

so that

$$\frac{\partial x_T}{\partial \mu_y} = e^{\mu_y + t\sigma_y} \tag{6-44}$$

$$\frac{\partial x_T}{\partial \sigma_y^2} = \frac{t\, e^{\mu_y + t\sigma_y}}{2\sigma_y} \tag{6-45}$$

so that

$$S_T^2 = \omega^2 \text{ var } \mu_y + \frac{t^2\omega^2}{4\sigma_y^2} \text{ var } \sigma_y^2 + \frac{t\omega^2}{\sigma_y} \text{ cov } (\mu_y, \sigma_y^2) \tag{6-46}$$

where $\omega$ represents $e^{\mu_y + t\sigma_y}$. The logarithm of the likelihood equation (Equation 6-19) is

$$\ln L = -\frac{n}{2}\ln 2\pi - \frac{n}{2}\ln \sigma_y^2 - \sum_{i=1}^{n} \ln x_i - \sum_{i=1}^{n} (\ln x_i - \mu_y)^2/2\sigma_y^2 \tag{6-47}$$

so that

$$\frac{\partial \ln L}{\partial \mu_y} = \frac{\sum_{i=1}^{n} (\ln x_i - \mu_y)}{\sigma_y^2} \tag{6-48}$$

$$\frac{\partial \ln L}{\partial \sigma_y^2} = -\frac{n}{2\sigma_y^2} + \frac{\sum_{i=1}^{n} (\ln x_i - \mu_y)^2}{2\sigma_y^4} \tag{6-49}$$

$$\frac{\partial^2 \ln L}{\partial \mu_y^2} = -\frac{n}{\sigma_y^2} \tag{6-50}$$

$$\frac{\partial^2 \ln L}{\partial (\sigma_y^2)^2} = \frac{n}{2\sigma_y^4} - \frac{\sum_{i=1}^{n} (\ln x_i - \mu_y)^2}{\sigma_y^6} = -\frac{n}{2\sigma_y^4} \tag{6-51}$$

$$\frac{\partial^2 \ln L}{\partial \mu_y \partial \sigma_y^2} = -\frac{\sum_{i=1}^{n} (\ln x_i - \mu_y)}{\sigma_y^4} = 0 \tag{6-52}$$

The information matrix is

$$I = \frac{n}{\sigma_y^2}\begin{bmatrix} 1 & 0 \\ 0 & \frac{1}{2\sigma_y^2} \end{bmatrix} \tag{6-53}$$

with a determinant, D, of

$$D = \frac{1}{2\sigma_y^2} \tag{6-54}$$

Inverting this matrix gives the variances and covariances as

$$\text{var } \mu_y = \sigma_y^2/n \tag{6-55}$$

$$\text{var } \sigma_y^2 = 2\sigma_y^4/n \tag{6-56}$$

$$\text{cov } (\mu_y, \sigma_y^2) = 0 \tag{6-57}$$

Substituting these expressions in 6-46 yields

$$S_T^2 = \frac{\sigma_y^2 e^{2(\mu_y + t\sigma_y)}}{n} \left[1 + t^2/2\right] \tag{6-58}$$

where $S_T$ is now in linear units. However,

$$e^{\mu_y + t\sigma_y} = x_T \tag{6-59}$$

and

$$x_T = \mu\ (1 + Kz) \tag{6-60}$$

where K, the frequency factor is given by Equation 6-29 and z is the coefficient of variation. Substituting these relationships in 6-58 gives

$$S_T^2 = \frac{\sigma^2}{n}\left\{\frac{[\ln\ (z^2 + 1)](1 + Kz)^2\ (1 + t^2/2)}{z^2}\right\} \tag{6-61}$$

Simplifying to

$$S_T = \delta_z\sigma/\sqrt{n} \tag{6-62}$$

Table 6-4 gives values of $\delta_z$ for some values of z and t.

Table 6-4

Parameter $\delta_Z$ for Use in Standard Error of Lognormal Distribution

| Coefficient of Variation | Cumulative Probability, P, % 50 | 80 | 90 | 95 | 98 | 99 |
|---|---|---|---|---|---|---|
| Z | Corresponding Return Period, T, Years 2 | 5 | 10 | 20 | 50 | 100 |
| .05 | .9981 | 1.2114 | 1.4361 | 1.6622 | 1.9502 | 2.1584 |
| .10 | .9926 | 1.2562 | 1.5222 | 1.7940 | 2.1481 | 2.4099 |
| .15 | .9834 | 1.2975 | 1.6068 | 1.9280 | 2.3557 | 2.6786 |
| .20 | .9710 | 1.3348 | 1.6890 | 2.0629 | 2.5714 | 2.9632 |
| .25 | .9555 | 1.3679 | 1.7679 | 2.1974 | 2.7935 | 3.2617 |
| .30 | .9373 | 1.3963 | 1.8426 | 2.3301 | 3.0201 | 3.5720 |
| .35 | .9167 | 1.4201 | 1.9126 | 2.4597 | 3.2491 | 3.8918 |
| .40 | .8942 | 1.4391 | 1.9773 | 2.5850 | 3.4786 | 4.2184 |
| .45 | .8702 | 1.4535 | 2.0362 | 2.7052 | 3.7065 | 4.5493 |
| .50 | .8450 | 1.4634 | 2.0892 | 2.8192 | 3.9312 | 4.8819 |
| .55 | .8190 | 1.4690 | 2.1360 | 2.9264 | 4.1509 | 5.2137 |
| .60 | .7925 | 1.4706 | 2.1768 | 3.0263 | 4.3642 | 5.5424 |
| .65 | .7658 | 1.4686 | 2.2115 | 3.1187 | 4.5699 | 5.8659 |
| .70 | .7390 | 1.4632 | 2.2405 | 3.2032 | 4.7670 | 6.1823 |
| .75 | .7126 | 1.4549 | 2.2639 | 3.2799 | 4.9547 | 6.4901 |
| .80 | .6865 | 1.4440 | 2.2820 | 3.3489 | 5.1324 | 6.7879 |
| .85 | .6610 | 1.4308 | 2.2953 | 3.4103 | 5.2998 | 7.0747 |
| .90 | .6362 | 1.4156 | 2.3040 | 3.4644 | 5.4567 | 7.3496 |
| .95 | .6120 | 1.3987 | 2.3085 | 3.5114 | 5.6030 | 7.6121 |
| 1.00 | .5887 | 1.3805 | 2.3093 | 3.5518 | 5.7387 | 7.8618 |

## Computer Program

The following pages list, in sequence, a sample data set, a computer program, and the corresponding output. The program fits a two-parameter lognormal distribution to the data set by the method of moments and by the method of maximum likelihood as described in the preceeding sections. The program also computes the T-year event magnitudes and standard errors at return periods of 2, 5, 10, 20, 50 and 100 years using parameters derived by both moments and maximum likelihood.

The program is written in FORTRAN IV suitable for use on a CDC 6400 digital computer.

Subsequent chapters also list FORTRAN programs and, for each distribution, the output listed corresponds to the data sample given on the next page.

While each of the programs has been tested on several sets of data it is possible that errors exist or that there are limitations on the applicability of some or any of the programs. Use of non-original computer programs always involves some risk. The programs have, in each case, been written as simply as possible- and may not give solutions for data sets having multiple roots. An alternate computer program designed to accept multiple roots has been written and tested by Condie and Nix (4). This alternate program fits the 2- and 3-parameter lognormal distributions, the type I extremal distribution and the log-Pearson type III distribution.

## References


1. Chow, V. T., 1954, The Log-Probability Law and its Engineering Applications, Proc. ASCE, Vol. 80, pp. 1-25.

2. Chow, V. T., 1964, Editor-in-Chief, Handbook of Applied Hydrology, McGraw-Hill.

3. Collier, E. P., 1965, Flood Frequency Curves - Single Station Analysis, Water Resources Branch, Dept. Environment, Ottawa.

4. Condie, R. and G. Nix, 1976, Flood Damage Reduction Frequency Analysis Program, Inland Waters Directorate, Dept. Environment, Ottawa.

5. Hardison, C. H., 1969, Accuracy of Streamflow Characteristics, USGS Professional Paper No. 650-D, pp. D210-D214.

6. Jennings, M. E. and M. A. Benson, 1969, Frequency Curves for Annual Flood Series with Some Zero Events or Incomplete Data, Wat. Res. Res., Vol. 5, No. 1, pp. 276-280.

7. Kilmartin, R. F. and J. R. Peterson, 1972, Rainfall-Runoff Regression with Logarithmic Transforms and Zeros in the Data, Wat. Res. Res., Vol. 8, No. 4, pp. 1096-1099.

8. Sangal, B. P. and A. K. Biswas, 1970, The 3-Parameter Lognormal Distribution and its Applications in Hydrology, Wat. Res. Res., Vol. 6, No. 2, pp. 505-515.

9. Singh, K. P. and R. A. Sinclair, 1972, Two-Distribution Method for Flood-Frequency Analysis, Proc. ASCE, Vol. 98, No. HY1, pp. 29-45.

10. WMO, 1969, Estimation of Maximum Floods, Technical Note No. 98, WMO No. 233.TP.126, Geneva.

## SAMPLE DATA

### ST. MARYS RIVER AT STILLWATER, NOVA SCOTIA
### STATION NO. 01E0001

| YEAR | ANNUAL MAXIMUM DAILY DISCHARGE* | ANNUAL MINIMUM DAILY DISCHARGE** |
|---|---|---|
| 1915 | --- | 143 CFS ON SEP 21 |
| 1916 | 10400 CFS ON APR 7 | 37.0 CFS ON AUG 26 |
| 1917 | 10700 CFS ON APR 8 | 37.0 CFS ON SEP 17 |
| 1918 | 20100 CFS ON JAN 5 | 89.0 CFS ON AUG 25 |
| 1919 | 8210 CFS ON APR 19 | 109 CFS ON JUL 10 |
| 1920 | 14300 CFS ON MAR 15 | 109 CFS ON JUL 31 |
| 1921 | 8040 CFS ON MAR 12 | 23.0 CFS ON SEP 22 |
| 1922 | 8210 CFS ON JUL 6 | 98.0 CFS ON JUN 19 |
| 1923 | 13900 CFS ON OCT 3 | 71.0 CFS ON MAR 16 |
| 1924 | 8390 CFS ON JUN 17 | 56.0 CFS ON AUG 5 |
| 1925 | 18500 CFS ON APR 1 | 43.0 CFS ON SEP 8 |
| 1926 | 13000 CFS ON DEC 2 | 37.0 CFS ON SEP 20 |
| 1927 | 16400 CFS ON OCT 20 | 237 CFS ON SEP 18 |
| 1928 | 14500 CFS ON DEC 11 | 63.0 CFS ON SEP 12 |
| 1929 | 13000 CFS ON SEP 20 | 79.0 CFS ON JUL 14 |
| 1930 | 17200 CFS ON MAR 13 | 40.5 CFS ON SEP 14 |
| 1931 | 13900 CFS ON APR 3 | 97.0 CFS ON AUG 13 |
| 1932 | 11900 CFS ON DEC 2 | 60.0 CFS ON JUL 21 |
| 1933 | 13600 CFS ON OCT 9 | 56.0 CFS ON AUG 11 |
| 1934 | 12400 CFS ON APR 15 | 16.0 CFS ON SEP 15 |
| 1935 | 18300 CFS ON JAN 12 | 36.8 CFS ON AUG 22 |
| 1936 | 12900 CFS ON MAR 14 | 122 CFS ON AUG 6 |
| 1937 | 18200 CFS ON NOV 21 | 19.9 CFS ON SEP 8 |
| 1938 | 9900 CFS ON NOV 26 | 140 CFS ON JUN 27 |
| 1939 | 10200 CFS ON APR 21 | 23.8 CFS ON SEP 6 |
| 1940 | 9020 CFS ON DEC 10 | 36.8 CFS ON AUG 19 |
| 1941 | 11800 CFS ON APR 29 | 79.0 CFS ON JUL 8 |
| 1942 | 16100 CFS ON OCT 27 | 5.3 CFS ON SEP 9 |
| 1943 | 16900 CFS ON MAY 5 | 107 CFS ON FEB 3 |
| 1944 | 11800 CFS ON NOV 7 | 20.8 CFS ON AUG 25 |
| 1945 | 13900 CFS ON JAN 18 | 45.3 CFS ON AUG 21 |
| 1946 | 12300 CFS ON DEC 23 | 20.8 CFS ON JUL 21 |
| 1947 | 15100 CFS ON MAY 3 | 16.0 CFS ON AUG 24 |
| 1948 | 11900 CFS ON APR 3 | 118 CFS ON AUG 6 |
| 1949 | 11000 CFS ON NOV 27 | 63.0 CFS ON AUG 3 |
| 1950 | 16000 CFS ON NOV 29 | 14.3 CFS ON OCT 12 |
| 1951 | 11600 CFS ON MAY 30 | 74.0 CFS ON AUG 9 |
| 1952 | 19900 CFS ON JAN 24 | 25.7 CFS ON SEP 17 |
| 1953 | 18600 CFS ON FEB 9 | 25.7 CFS ON AUG 4 |
| 1954 | 18000 CFS ON FEB 25 | 40.5 CFS ON SEP 30 |
| 1955 | 13100 CFS ON FEB 13 | 53.0 CFS ON AUG 18 |
| 1956 | 29100 CFS ON JAN 7 | 41.3 CFS ON OCT 6 |
| 1957 | 10300 CFS ON NOV 4 | 35.6 CFS ON SEP 2 |
| 1958 | 12200 CFS ON NOV 30 | 83.0 CFS ON JUL 29 |
| 1959 | 15600 CFS ON APR 5 | 260 CFS ON JAN 16 |
| 1960 | 12700 CFS ON APR 7 | 7.8 CFS ON SEP 12 |
| 1961 | 13100 CFS ON DEC 30 | 19.1 CFS ON AUG 26 |
| 1962 | 19200 CFS ON APR 3 | 192 CFS ON JUL 1 |
| 1963 | 19500 CFS ON MAY 3 | 119 CFS ON AUG 13 |
| 1964 | 23000 CFS ON APR 17 | 270 CFS ON AUG 23 |
| 1965 | 6700 CFS ON APR 18 | 69.3 CFS ON OCT 1 |
| 1966 | 7130 CFS ON APR 2 | 34.0 CFS ON SEP 4 |
| 1967 | 14300 CFS ON NOV 25 | 132 CFS ON JUL 30 |
| 1968 | 20600 CFS ON AUG 31 | 17.9 CFS ON AUG 20 |
| 1969 | 25600 CFS ON NOV 10 | 29.6 CFS ON SEP 5 |
| 1970 | 8180 CFS ON JUL 13 | 175 CFS ON JAN 28 |
| 1971 | 34400 CFS ON AUG 16 | 82.8 CFS ON JUL 14 |
| 1972 | 16100 CFS ON MAY 17 | 227 CFS ON FEB 13 |
| 1973 | 10200 CFS ON APR 30 | 67.9 CFS ON OCT 4 |
| 1974 | 12300 CFS ON DEC 4 | 42.8 CFS ON AUG 18 |

* Data used for 2- and 3-parameter lognormal, type 1 extremal, Pearson type III and log-Pearson type III distributions.

** Data used for type III extremal distribution.

```
      PROGRAM LN2(INPUT,OUTPUT,TAPE5=INPUT,TAPE6=OUTPUT)
C
C
C     COMPUTES METHOD OF MOMENTS AND MAXIMUM LIKELIHOOD ESTIMATES FOR
C     T YEAR EVENTS AND STANDARD ERRORS FOR 2 PARAMETER LOGNORMAL
C     DISTRIBUTION
C     INPUT
C     TITLE
C     N NUMBER OF ANNUAL MAXIMUM EVENTS
C     X SERIES OF EVENTS
C
C
      DIMENSION SND(6), X(100)
      DIMENSION XT(6), SX(6), TITLE(80)
      REAL K,M1,M2,M3
      SND(1)=0.0
      SND(2)=0.8416
      SND(3)=1.2816
      SND(4)=1.6449
      SND(5)=2.0538
      SND(6)=2.3264
      READ (5,8) TITLE
      READ (5,9) N
      XN=N
      READ (5,10) (X(I),I=1,N)
      WRITE (6,11) TITLE
      WRITE (6,12)
      A=0.0
      B=0.0
      C=0.0
      DO 1 I=1,N
      A=A+X(I)
      B=B+X(I)**2
      C=C+X(I)**3
1     CONTINUE
      M1=A/XN
      M2=(B/XN)-(A/XN)**2
      M3=(C/XN)+2.0*M1**3-3.0*M1*(B/XN)
      M2=M2*XN/(XN-1.0)
      G=M3/(M2**1.5)
      WRITE (6,5) M1
      WRITE (6,6) M2
      WRITE (6,7) G
      Z=(SQRT(M2))/M1
      A=ALOG(1.0+Z**2)
      DO 2 J=1,6
      T=SND(J)
      K=(EXP(SQRT(A)*T-A/2.0)-1.0)/Z
      XT(J)=M1+K*SQRT(M2)
      DELTA=SQRT(1.0+((Z**3+3.0*Z)*K)+((Z**8+6.0*Z**6+15.0*Z**4+16.0*Z**
     12+2.0)*K**2)/4.0)
2     SX(J)=DELTA*SQRT(M2/XN)
      WRITE (6,13)
      WRITE (6,14) (XT(J),J=1,6)
      WRITE (6,15) (SX(J),J=1,6)
      DO 3 J=1,6
      T=SND(J)
      K=(EXP(SQRT(A)*T-A/2.0)-1.0)/Z
      DELTA=SQRT((A*((1.0+K*Z)**2)*(1.0+(T**2)/2.0))/Z**2)
3     SX(J)=DELTA*SQRT(M2/XN)
      A=0.0
      B=0.0
      C=0.0
      DO 4 I=1,N
      X(I)=ALOG(X(I))
      A=A+X(I)
      B=B+X(I)**2
      C=C+X(I)**3
4     CONTINUE
      M1=A/XN
      M2=(B/XN)-(A/XN)**2
      M3=(C/XN)+2.0*M1**3-3.0*M1*(B/XN)
      M2=M2*XN/(XN-1.0)
      G=M3/(M2**1.5)
      WRITE (6,17) M1
      WRITE (6,18) M2
      WRITE (6,19) G
      WRITE (6,16)
      WRITE (6,13)
      WRITE (6,14) (XT(J),J=1,6)
      WRITE (6,15) (SX(J),J=1,6)
      STOP
```

```
C
5     FORMAT (20X,9HMEAN OF X,16X,E12.5)
6     FORMAT (20X,13HVARIANCE OF X,12X,E12.5)
7     FORMAT (20X,9HSKEW OF X,16X,E12.5,/)
8     FORMAT (80A1)
9     FORMAT (I5)
10    FORMAT (8F10.0)
11    FORMAT (1H1,/,80A1,//,21X,38H  TWO PARAMETER LOGNORMAL DISTRIBUTIO
     1N,/)
12    FORMAT (31X,17HMETHOD OF MOMENTS,//)
13    FORMAT (3X,7HT,YEARS,4X,1H2,11X,1H5,10X,2H10,10X,2H20,10X,2H50,9X,
     13H100,/)
14    FORMAT (3X,1HX,3X,6E12.5,/,4X,1HT)
15    FORMAT (3X,1HS,3X,6E12.5,/,4X,1HT,//)
16    FORMAT (3X,77HNOTE - FOR GOOD USE OF THIS DISTRIBUTION SKEW OF LOG
     1S SHOULD BE CLOSE TO ZERO,/)
17    FORMAT (20X,15HMEAN OF LN(X)  ,10X,E12.5)
18    FORMAT (20X,19HVARIANCE OF LN(X)  ,6X,E12.5)
19    FORMAT (20X,15HSKEW OF LN(X)  ,10X,E12.5,/)
      END
```

```
              ST. MARYS RIVER AT STILLWATER - STATION NO. 01E0001

                      TWO PARAMETER LOGNORMAL DISTRIBUTION

                               METHOD OF MOMENTS


                    MEAN OF X                         .14555E+05
                    VARIANCE OF X                     .27320E+08
                    SKEW OF X                         .12887E+01

T,YEARS    2           5           10          20          50          100

X      .13698E+05  .18364E+05  .21405E+05  .24292E+05  .28010E+05  .30800E+05
 T
S      .62026E+03  .10439E+04  .14042E+04  .17657E+04  .22441E+04  .26082E+04
 T


                         MAXIMUM LIKELIHOOD PROCEDURE

                    MEAN OF LN(X)                     .95285E+01
                    VARIANCE OF LN(X)                 .11384E+00
                    SKEW OF LN(X)                     .18649E+00

NOTE - FOR GOOD USE OF THIS DISTRIBUTION SKEW OF LOGS SHOULD BE CLOSE TO ZERO

T,YEARS    2           5           10          20          50          100

X      .13698E+05  .18364E+05  .21405E+05  .24292E+05  .28010E+05  .30800E+05
 T
S      .61592E+03  .96085E+03  .12989E+04  .16754E+04  .22207E+04  .26661E+04
 T
```

## CHAPTER 7

## THREE-PARAMETER LOGNORMAL DISTRIBUTION

### Introduction

Just as the lognormal distribution represents the normal distribution of the logarithms of the variable x, so the 3- parameter lognormal represents the normal distribution of the logarithms of the reduced variable (x-a) where a is a lower boundary. The probability density distribution is then given by:

$$p(x) = \frac{1}{(x-a)\,\sigma_y\sqrt{2\pi}}\, e^{-\frac{[\ln(x-a)-\mu_y]^2}{2\sigma_y^2}} \qquad (7\text{-}1)$$

where $\mu_y$ and $\sigma_y^2$ are the form and scale parameters, shown later to be the mean and variance of the logarithms of (x-a).

### Estimation of Parameters

If the lower boundary, a, is known then the reduced variable (x-a) can be used together with the procedures described for the 2-parameter lognormal distribution. If a is unknown the following methods are among those available.

*Method of Moments* - Applying the general moment generating equation for moments about a, the lower bound, to the pdf for the 3-parameter lognormal distribution (Equation 7-1)

$$\mu_r(a) = \int_0^\infty \frac{(x-a)^r}{(x-a)\,\sigma_y\sqrt{2\pi}} \exp\left\{\left[-\ln(x-a) - \mu_y\right]^2 / 2\sigma_y^2\right\} dx \qquad (7\text{-}2)$$

Substitute y = ln (x-a), $(x-a)^r = e^{ry}$, dx = (x-a) dy so that

$$\mu_r(a) = \frac{1}{\sigma_y\sqrt{2\pi}} \int_{-\infty}^{\infty} e^{ry}\, e^{-(y-\mu_y)^2/2\sigma_y^2}\, dx \qquad (7\text{-}3)$$

Complete the square within the exponents as

$$-\left\{(y - (\mu_y + r\sigma_y^2))^2 - r^2\sigma_y^4 - 2r\mu_y\sigma_y^2\right\}/2\sigma_y^2 \qquad (7\text{-}4)$$

then

$$\mu_r(a) = e^{r\mu_y + r^2\sigma_y^2/2} \frac{1}{\sigma_y\sqrt{2\pi}} \int_{-\infty}^{\infty} e^{-[y - (\mu_y + r\sigma_y^2)]^2/2\sigma_y^2}\, dy \qquad (7\text{-}5)$$

Since this integral is unity

$$\mu_r(a) = e^{r\mu_y + r^2\sigma_y^2/2} \tag{7-6}$$

The moments about the lower bound, a, can now be converted to moments about the origin by using the equation

$$\mu'_r = \sum_{j=0}^{r} \binom{r}{j} \mu_{r-j}(a) \cdot a^j \tag{7-7}$$

so that

$$\mu'_1 = \mu = a + e^{\mu_y + \sigma_y^2/2} \tag{7-8}$$

Finally, converting to central moments leads to

$$\mu_2 = \sigma^2 = \left(e^{\sigma_y^2} - 1\right) e^{2\mu_y + \sigma_y^2} \tag{7-9}$$

etc. If $z_1$ and $z_2$ represent the coefficients of variation of the distributions x and x-a then

$$z_1 = \sigma/\mu \tag{7-10}$$

and

$$z_2 = \sigma/(\mu - a) \tag{7-11}$$

from which

$$a = \mu\ (1 - z_1/z_2) = \mu - \sigma/z_2 \tag{7-12}$$

The value of $z_1$ can be computed directly from the observed events. Since the second and third moments of the distribution (x-a) do not contain terms in a, the value of $z_2$ can be obtained from the relationship

$$\gamma_1 = 3z_2 + z_2^3 \tag{7-13}$$

where $\gamma_1$ is the coefficient of skew of the distribution x. The solution of this equation is

$$z_2 = \frac{1 - \omega^{2/3}}{\omega^{1/3}} \tag{7-14}$$

where

$$\omega = \frac{-\gamma_1 + (\gamma_1^2 + 4)^{\frac{1}{2}}}{2} \tag{7-15}$$

The procedure is then to compute the mean, $\mu$, standard deviation, $\sigma$, and coefficient of skew, $\gamma_1$, of the observed events, x; compute $z_1$ directly and $z_2$ from 7-14; compute a from Equation 7-12 and then

$$\sigma_y = [\ln (z_2^2 + 1)]^{\frac{1}{2}} \tag{7-16}$$

and

$$\mu_y = \ln (\sigma/z_2) - \tfrac{1}{2} \ln (z_2^2 + 1) \tag{7-17}$$

*Maximum Likelihood* - Taking the product of Equation 7-1 for $x_1 \rightarrow x_n$ and taking logarithms yields the logarithmic likelihood function:

$$\ln L = - \sum_{i=1}^{n} \ln (x_i - a) - \frac{n}{2} \ln (2\pi) - n \ln \sigma_y$$

$$- \frac{\sum_{i=1}^{n} [\ln (x_i - a) - \mu_y]^2}{2\sigma_y^2} \tag{7-18}$$

Differentiating with respect to $\mu_y$, $\sigma_y^2$ and a and equating to zero leads to the three expressions:

$$\mu_y = \sum_{i=1}^{n} \ln (x_i - a)/n \tag{7-19}$$

$$\sigma_y^2 = \sum_{i=1}^{n} [\ln (x_i - a) - \mu_y]^2/n \tag{7-20}$$

and

$$\sum_{i=1}^{n} (x_i - a)^{-1} (\mu_y - \sigma_y^2) = \sum_{i=1}^{n} [(1/(x_i - a)) \ln (x_i - a)] \tag{7-21}$$

By substituting for $\mu_y$ and $\sigma_y^2$ in Equation 7-21 an equation in a only is found which can be solved numerically.

*Other Methods* - Computation of the coefficient of skew from small samples is sometimes subject to error. Sangal and Biswas (3) derived a method of estimating the parameter a using only the mean, median and standard deviation of the original data. Their solution is:

$$a = \delta - \frac{\sigma^2}{2(\mu - \delta)} \tag{7-22}$$

where $\delta$ is the median of x, determined as the mean of the middle 1/5th of the data. The determination of a is very sensitive to the difference between the mean and median, $(\mu - \delta)$. When this difference becomes small, then a takes a large negative value.

The assumption in this case is that the original data are symmetrically, but not necessarily normally, distributed. This follows from the empirical relationship for moderately skewed distributions that (Mean - Mode) = 3(Mean - Median) so that if the median equals the mean then the mode also equals the mean, which only occurs in symmetrical distributions. In such a case if the median is set to 0.99 $\mu$ then equation 7-22 can be modified to

$$a = \mu\,(0.99 - 50\,z^2) \qquad (7\text{-}23)$$

where z is the coefficient of variation. Burges et al. (1) compared the performance of the methods of moments and median and concluded that moments was better.

Condie (2) has described a simple graphical method of determining the parameter a which is applicable provided that:

(a) At least one of the graph scales, vertical or horizontal, is logarithmic.
(b) The curvature of the best fitting line is gradually decreasing.

The other scale can be linear, logarithmic, normally probabilistic, time scale, etc.

Referring to Figure 7-1; $(x_1,y_1)$, $(x_2,y_2)$, $(x_3,y_3)$ are three

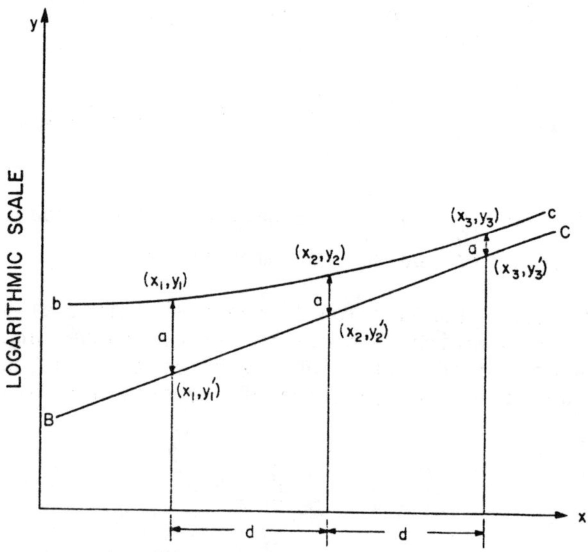


Figure 7-1. Graphical technique of estimating parameter a in 3-parameter lognormal distribution.

points on the best-fitting line through the plotted event magnitudes (b-c) such that

$$x_3 - x_1 = x_2 - x_3 \tag{7-24}$$

The object is then to find a constant a, which will move points $(x_1,y_1)$, $(x_2,y_2)$, $(x_3,y_3)$ to points $(x_1,y_1')$, $(x_2,y_2')$, $(x_3,y_3')$ all on a straight line (B-C). The straight line is easier to extrapolate than the original arbitrarily curved line.

Considering the slope of the logarithmically straight line B-C:

$$\frac{\ln(y_3-a) - \ln(y_1-a)}{x_3 - x_1} = \frac{\ln(y_2-a) - \ln(y_3-a)}{x_2 - x_1} \tag{7-25}$$

but from Equation 7-24 this reduces to

$$\ln\frac{y_3-a}{y_1-a} = \ln\frac{y_2-a}{y_3-a} \tag{7-26}$$

from which

$$a = \frac{y_1y_2-y_3^2}{y_1+y_2-2y_3} \tag{7-27}$$

The method is simple and can be applied to any distribution with a logarithmic scale such as lognormal, log-Gumbel, log-Pearson type III.

Frequency Factor

Using the standard normal deviate as frequency factor the first expression obtained is

$$y_T = \mu_y + t\sigma_y \tag{7-28}$$

where $\mu_y$ and $\sigma_y$ are the mean and standard deviation of the series ln (x-a) so that the T-year event, $x_T$, is

$$x_T = a + e^{\mu_y + t\sigma_y} \tag{7-29}$$

Secondly, using the standard frequency equation

$$x_T = \mu + K\sigma \tag{7-30}$$

the frequency factor K can be found by substituting for $x_T$, $\mu$ and $\sigma$ as

$$K = \frac{\exp\,(y_T - \mu_y - \sigma_y^2/2) - 1}{(e^{\sigma_y^2} - 1)^{\frac{1}{2}}} \qquad (7\text{-}31)$$

Now, substituting for $y_T$ and $\mu_y$ in terms of $\sigma_y$ from 7-28 and substituting for $\sigma_y$ in terms of $z_2$ from 7-16 yields:

$$K = \frac{\exp\,\{[\ln\,(1 + z_2^2)]^{\frac{1}{2}} t - [\ln\,(1 + z_2^2)]/2\} - 1.0}{z_2} \qquad (7\text{-}32)$$

But from Equations 7-14 and 7-15, $z_2$ is a function of $\gamma_1$, the sample coefficient of skew. Table 7-1 lists values of K for some typical values of $\gamma_1$. Note that at $\gamma_1 = 0.0$ the frequency factor K is zero for all return periods. The 3-parameter lognormal (together with the 2-parameter lognormal) is thus not a suitable distribution for use with data samples having skews approaching zero.

## Standard Error

*Method of Moments* - As with the two-parameter lognormal distribution, 2 options are available to compute the standard error. The first method, using the frequency equation

Table 7-1

Frequency Factor for use in 3-Parameter Lognormal Distribution

| Coefficient of skew $\gamma_1$ | Cumulative Probability, P, % | | | | | |
|---|---|---|---|---|---|---|
| | 50 | 80 | 90 | 95 | 98 | 99 |
| | Corresponding Return Period, T, Years | | | | | |
| | 2 | 5 | 10 | 20 | 50 | 100 |
| -2.00 | .2366 | -.6144 | -1.2437 | -1.8916 | -2.7943 | -3.5196 |
| -1.80 | .2240 | -.6395 | -1.2621 | -1.8928 | -2.7578 | -3.4433 |
| -1.60 | .2092 | -.6654 | -1.2792 | -1.8901 | -2.7138 | -3.3570 |
| -1.40 | .1920 | -.6920 | -1.2943 | -1.8827 | -2.6615 | -3.2601 |
| -1.20 | .1722 | -.7186 | -1.3067 | -1.8696 | -2.6002 | -3.1521 |
| -1.00 | .1495 | -.7449 | -1.3156 | -1.8501 | -2.5294 | -3.0333 |
| - .80 | .1241 | -.7700 | -1.3201 | -1.8235 | -2.4492 | -2.9043 |
| - .60 | .0959 | -.7930 | -1.3194 | -1.7894 | -2.3600 | -2.7665 |
| - .40 | .0654 | -.8131 | -1.3128 | -1.7478 | -2.2631 | -2.6223 |
| - .20 | .0332 | -.8296 | -1.3002 | -1.6993 | -2.1602 | -2.4745 |
| .00 | 0.0000 | 0.0000 | 0.0000 | 0.0000 | 0.0000 | 0.0000 |
| .20 | -.0332 | .8296 | 1.3002 | 1.6993 | 2.1602 | 2.4745 |
| .40 | -.0654 | .8131 | 1.3128 | 1.7478 | 2.2631 | 2.6223 |
| .60 | -.0959 | .7930 | 1.3194 | 1.7894 | 2.3600 | 2.7665 |
| .80 | -.1241 | .7700 | 1.3201 | 1.8235 | 2.4492 | 2.9043 |
| 1.00 | -.1495 | .7449 | 1.3156 | 1.8501 | 2.5294 | 3.0333 |
| 1.20 | -.1722 | .7186 | 1.3067 | 1.8696 | 2.6002 | 3.1521 |
| 1.40 | -.1920 | .6920 | 1.2943 | 1.8827 | 2.6615 | 3.2601 |
| 1.60 | -.2092 | .6654 | 1.2792 | 1.8901 | 2.7138 | 3.3570 |
| 1.80 | -.2240 | .6395 | 1.2621 | 1.8928 | 2.7578 | 3.4433 |
| 2.00 | -.2366 | .6144 | 1.2437 | 1.8916 | 2.7943 | 3.5196 |

$$y_T = \ln (x_T - a) = \mu_y + t\sigma_y \tag{7-33}$$

gives

$$S_T^2 = \frac{\sigma_y^2}{n} \left\{1 + t^2/2\right\} \tag{7-34}$$

but $S_T^2$ is now in logarithmic units and refers to $\ln (x_T - a)$ not to $x_T$.

Secondly we can use the frequency equation

$$x_T = \mu + K\sigma \tag{7-35}$$

where $\mu$ and $\sigma$ are the mean and standard deviation of $x_i$, i = 1, n and K is the frequency factor from Section 7-3.

Since K is a function of $\gamma_1$ and T, Equation 7-35 can be rewritten as

$$x_T = f(\mu_1', \mu_2, \mu_3, T) \tag{7-36}$$

and since the return period, T, is invariable

$$\begin{aligned} \text{var } x_T = {} & \left(\frac{\partial x_T}{\partial \mu_1'}\right)^2 \text{var } \mu_1' + \left(\frac{\partial x_T}{\partial \mu_2}\right)^2 \text{var } \mu_2 + \left(\frac{\partial x_T}{\partial \mu_3}\right)^2 \text{var } \mu_3 \\ & + 2 \frac{\partial x_T}{\partial \mu_1'} \frac{\partial x_T}{\partial \mu_2} \text{cov } (\mu_1', \mu_2) + 2 \frac{\partial x_T}{\partial \mu_1'} \frac{\partial x_T}{\partial \mu_3} \text{cov } (\mu_1', \mu_3) \\ & + 2 \frac{\partial x_T}{\partial \mu_2} \frac{\partial x_T}{\partial \mu_3} \text{cov } (\mu_2, \mu_3) \end{aligned} \tag{7-37}$$

where, from Equation 7-35

$$\frac{\partial x_T}{\partial \mu_1'} = 1 \tag{7-38}$$

$$\frac{\partial x_T}{\partial \mu_2} = \frac{1}{2\sigma} \left[K - 3\gamma_1 \frac{\partial K}{\partial \gamma_1}\right] \tag{7-39}$$

$$\frac{\partial x_T}{\partial \mu_3} = \frac{1}{\sigma^2} \frac{\partial K}{\partial \gamma_1} \tag{7-40}$$

From Equations 7-14, 7-15 and 7-32

$$\frac{\partial K}{\partial \gamma_1} = \frac{\partial K}{\partial z_2} \cdot \frac{\partial z_2}{\partial \omega} \cdot \frac{\partial \omega}{\partial \gamma_1} \qquad (7\text{-}41)$$

with

$$\frac{\partial K}{\partial z_2} = \left\{\frac{2z_2}{1 + z_2^2}\right\}\left\{\left[\exp \{[\ln (1 + z_2^2)]^{\frac{1}{2}} t - [\ln (1 + z_2^2)]/2\}\right].\right.$$

$$\left.\left[\frac{t}{2z_2 [\ln (1 + z_2^2)]^{\frac{1}{2}}} - \frac{1}{z_2} - \frac{1}{2z_2^3}\right] + \frac{1}{2z_2^3} + \frac{1}{2z_2}\right\} \qquad (7\text{-}42)$$

$$\frac{\partial z_2}{\partial \omega} = -\frac{1}{3\omega^{2/3}}\left[1 + \frac{1}{\omega^{2/3}}\right] \qquad (7\text{-}43)$$

$$\frac{\partial \omega}{\partial \gamma_1} = -\frac{1}{2} + \frac{\gamma_1}{2(\gamma_1^2 + 4)^{\frac{1}{2}}} \qquad (7\text{-}44)$$

Also, from Standard Error by Moments in Chapter 3,

$$\text{var } \mu_1' = \mu_2/n \qquad (7\text{-}45)$$

$$\text{var } \mu_2 = \frac{1}{n}\left\{\mu_4 - \mu_2^2\right\} \qquad (7\text{-}46)$$

$$\text{var } \mu_3 = \frac{1}{n}\left\{\mu_6 - \mu_3^2 - 6\mu_4\mu_2 + 9\mu_2^3\right\} \qquad (7\text{-}47)$$

$$\text{cov } (\mu_1', \mu_2) = \mu_3/n \qquad (7\text{-}48)$$

$$\text{cov } (\mu_1', \mu_3) = \frac{1}{n}\left\{\mu_4 - 3\mu_2^2\right\} \qquad (7\text{-}49)$$

$$\text{cov } (\mu_2, \mu_3) = \frac{1}{n}\left\{\mu_5 - 4\mu_3\mu_2\right\} \qquad (7\text{-}50)$$

and central moments may be obtained from Method of Moments in this chapter as

$$\mu_2 = e^{\sigma_y^2 + 2\mu_y}\left(e^{\sigma_y^2} - 1\right) \qquad (7\text{-}51)$$

$$\mu_3 = e^{3\sigma_y^2/2 + 3\mu_y}\left(e^{\sigma_y^2} - 1\right)^2\left(e^{\sigma_y^2} + 2\right) \qquad (7\text{-}52)$$

$$\mu_4 = e^{2\sigma_y^2 + 4\mu_y}\left(e^{\sigma_y^2} - 1\right)^2\left(e^{4\sigma_y^2} + 2e^{3\sigma_y^2} + 3e^{2\sigma_y^2} - 3\right) \quad (7\text{-}53)$$

$$\mu_5 = e^{5\sigma_y^2/2 + 5\mu_y}\left(e^{10\sigma_y^2} - 5e^{6\sigma_y^2} + 10e^{3\sigma_y^2} - 10e^{\sigma_y^2} + 4\right) \quad (7\text{-}54)$$

$$\mu_6 = e^{3\sigma_y^2 + 6\mu_y}\left(e^{15\sigma_y^2} - 6e^{10\sigma_y^2} + 15e^{6\sigma_y^2} - 20e^{3\sigma_y^2} + 15e^{\sigma_y^2} - 5\right) \quad (7\text{-}55)$$

Substitution of these results into Equation 7-37 ultimately yields an expression for the standard error of estimate.

The initial computations are to obtain the mean, variance and coefficient of skew, $\mu_1'$, $\mu_2$ and $\gamma_1$, from the observed events; calculate $\mu_y$ and $\sigma_y$ and the moments $\mu_4$ - $\mu_6$. The variances and covariances of the moments, Equation 7-45 - 7-50, and the derivatives $\partial x_T/\partial\mu_1'$, $\partial x_T/\mu_2$ and $\partial x_T/\partial\mu_3$ can then be calculated and the variance of the estimate obtained from 7-37.

*Maximum Likelihood* - For the 3-parameter lognormal distribution with pdf

$$f(x) = \frac{1}{(x-a)\,\sigma_y\sqrt{2\pi}}\exp\left\{-[(\ln x\text{-}a) - \mu_y]^2/2\sigma_y^2\right\} \quad (7\text{-}56)$$

the standard error of estimate by maximum likelihood is given by:

$$S_T^2 = \left(\frac{\partial x}{\partial a}\right)^2 \text{var } a + \left(\frac{\partial x}{\partial\sigma_y^2}\right)^2 \text{var } \sigma_y^2 + \left(\frac{\partial x}{\partial\mu_y}\right)^2 \text{var } \mu_y$$

$$+ 2\frac{\partial x}{\partial a}\frac{\partial x}{\partial\sigma_y^2}\text{cov }(a, \sigma_y^2) + 2\frac{\partial x}{\partial a}\frac{\partial x}{\partial\mu_y}\text{cov }(a, \mu_y)$$

$$+ 2\frac{\partial x}{\partial\sigma_y^2}\frac{\partial x}{\partial\mu_y}\text{cov }(\sigma_y^2, \mu_y) \quad (7\text{-}57)$$

From Equation 7-29

$$x_T = a + e^{\mu_y + t\sigma_y} \quad (7\text{-}58)$$

so that

$$\frac{\partial x}{\partial a} = 1 \quad (7\text{-}59)$$

$$\frac{\partial x}{\partial \sigma_y^2} = \frac{te^{\mu_y + t\sigma_y}}{2\sigma_y} \tag{7-60}$$

$$\frac{\partial x}{\partial \mu_y} = e^{\mu_y + t\sigma_y} \tag{7-61}$$

so that, writing $\omega$ for $e^{\mu_y + t\sigma_y}$ Equation 7-57 becomes

$$S_T^2 = \text{var } a + \frac{t^2\omega^2}{4\sigma_y^2} \text{ var } \sigma_y^2 + \omega^2 \text{ var } \mu_y + \frac{t\omega}{\sigma_y} \text{ cov } (a, \sigma_y^2)$$

$$+ 2\omega \text{ cov } (a, \mu_y) + \frac{t\omega^2}{\sigma_y} \text{ cov } (\sigma_y^2, \mu_y) \tag{7-62}$$

From Equation 7-56 the logarithm of the likelihood function is

$$\ln L = - \sum_{i=1}^{n} \ln (x_i - a) - \frac{n}{2} \ln \sigma_y^2 - \frac{n}{2} \ln (2\pi)$$

$$- \sum_{i=1}^{n} [\ln (x_i - a) - \mu_y]^2/2\sigma_y^2 \tag{7-63}$$

from which

$$\frac{\partial^2 \ln L}{\partial a^2} = \frac{1}{\sigma_y^2} \left\{(\sigma_y^2 - \mu_y - 1) \sum_{i=1}^{n} (x_i - a)^{-2} + \sum_{i=1}^{n} \left[\ln (x-a)\right]\left[x_i - a\right]^{-2}\right\} \tag{7-64}$$

$$\frac{\partial^2 \ln L}{\partial \mu_y^2} = - \frac{n}{\sigma_y^2} \tag{7-65}$$

$$\frac{\partial^2 \ln L}{\partial (\sigma_y^2)^2} = + \frac{n}{2\sigma_y^4} - \frac{\sum_{i=1}^{n} [\ln (x_i - a) - \mu_y]^2}{2\sigma_y^6} = \frac{n}{2\sigma_y^4} \tag{7-66}$$

$$\frac{\partial^2 \ln L}{\partial a \partial \mu_y} = - \frac{1}{\sigma_y^2} \sum_{i=1}^{n} (x_i - a)^{-1} \tag{7-67}$$

$$\frac{\partial^2 \ln L}{\partial a \partial(\sigma_y^2)} = - \frac{\Sigma[\ln (x-a) - \mu_y].(x-a)^{-1}}{\sigma_y^4} \qquad (7\text{-}68)$$

$$\frac{\partial^2 \ln L}{\partial \mu_y \partial(\sigma_y^2)} = - \frac{\sum_{i=1}^{n} [\ln (x-a) - \mu_y]}{\sigma_y^4} = 0 \qquad (7\text{-}69)$$

To evaluate $\sum_{i=1}^{n} (x_i - a)^r$, say M, introduce this factor into the pdf

$$M = \sum_{i=1}^{n} (x_i - a)^r = \frac{n}{\sigma_y \sqrt{2\pi}} \int_a^\infty \frac{(x_i - a)^r}{(x_i - a)}$$

$$\exp - \left\{ \left[ \ln (x_i - a) - \mu_y \right]^2 / 2\sigma_y^2 \right\} dx \qquad (7\text{-}70)$$

substitute $y = \ln (x_i - a)$

$$M = \frac{n}{\sigma_y \sqrt{2\pi}} \int_{-\infty}^{\infty} \exp \left\{ (- (y - \mu_y)^2 + 2ry\sigma_y^2)/2\sigma_y^2 \right\} dy \qquad (7\text{-}71)$$

completing the square within the exponent

$$M = e^{r\mu_y + r\sigma_y^2/2} \cdot \frac{n}{\sigma_y \sqrt{2\pi}} \int_{-\infty}^{\infty} \exp \left\{ - \left[ y - (\mu_y + r\sigma_y^2) \right]^2 / 2\sigma_y^2 \right\} dy$$

(7-72)

so that

$$\sum_{i=1}^{n} (x_i - a)^r = ne^{r\mu_y + r^2\sigma_y^2/2} \qquad (7\text{-}73)$$

Similarly, to evaluate $\sum_{i=1}^{n} \left\{ \left[ \ln (x_i - a) \right] \left[ x_i - a \right]^r \right\}$, say N,

$$N = \frac{n}{\sigma_y \sqrt{2\pi}} \int_a^\infty \frac{\ln(x_i - a)(x_i - a)^r}{(x_i - a)} \exp - \left\{ \left[ \ln (x_i - a) - \mu_y \right]^2 / 2\sigma_y^2 \right\} dx$$

(7-74)

Put $y = \ln (x_i - a)$, $x_i - a)^r = e^{ry}$, $dy = \frac{dx}{(x_i - a)}$, then

$$N = \frac{n}{\sigma_y \sqrt{2\pi}} \int_{-\infty}^{\infty} y \exp - \left\{ \left[ (y - \mu_y)^2 - 2ry\sigma_y^2 \right] / 2\sigma_y^2 \right\} dy \quad (7\text{-}75)$$

completing the square within the exponent

$$N = e^{r\mu_y + r^2\sigma_y^2/2} \frac{n}{\sigma_y \sqrt{2\pi}} \int_{-\infty}^{\infty} y\, e^{-\left[ y - (\mu_y + r\sigma_y^2) \right] / 2\sigma_y^2} dy \quad (7\text{-}76)$$

so that

$$\sum_{i=1}^{n} \left\{ \left[ \ln (x_i - a) \right] \left[ x_i - a \right]^r \right\} = n(\mu + r\sigma_y^2)\, e^{r\mu_y + r^2\sigma_y^2/2} ) \quad (7\text{-}77)$$

Utilizing these expressions the derivatives of the likelihood equation are simplified to

$$\frac{\partial^2 \ln L}{\partial a^2} = - \frac{n}{\sigma_y^2} \left\{ (1 + \sigma_y^2)\, e^{2\sigma_y^2 - 2\mu_y} \right\} \quad (7\text{-}78)$$

$$\frac{\partial^2 \ln L}{\partial \mu_y^2} = - \frac{n}{\sigma_y^2} \quad (7\text{-}79)$$

$$\frac{\partial^2 \ln L}{\partial (\sigma_y^2)^2} = - \frac{n}{2\sigma_y^4} \quad (7\text{-}80)$$

$$\frac{\partial^2 \ln L}{\partial a \partial \mu_y} = - \frac{n}{\sigma_y^2} e^{\sigma_y^2/2 - \mu_y} \quad (7\text{-}81)$$

$$\frac{\partial^2 \ln L}{\partial a \partial (\sigma_y^2)} = \frac{n}{\sigma_y^2} e^{\sigma_y^2/2 - \mu_y} \quad (7\text{-}82)$$

From these expressions, the likelihood or information matrix is:

$$I = \frac{n}{\sigma_y^2} \begin{bmatrix} (\sigma_y^2 + 1)\, e^{2(\sigma_y^2 - \mu_y)} & e^{\sigma_y^2/2 - \mu_y} & - e^{\sigma_y^2/2 - \mu_y} \\ e^{\sigma_y^2/2 - \mu_y} & 1 & 0 \\ - e^{\sigma_y^2/2 - \mu_y} & 0 & \frac{1}{2\sigma_y^2} \end{bmatrix} \tag{7-83}$$

The determinant, D, of this matrix is:

$$D = \frac{(\sigma_y^2 + 1)}{2\sigma_y^2} e^{2(\sigma_y^2 - \mu_y)} - \frac{(2\sigma_y^2 + 1)}{2\sigma_y^2} e^{\sigma_y^2 - 2\mu_y} \tag{7-84}$$

and, by inverting the matrix, the variances and covariances can be obtained as

$$\text{var } a = \frac{1}{2nD} \tag{7-85}$$

$$\text{var } \mu_y = \frac{\sigma_y^2}{nD} \left[ \frac{(\sigma_y^2 + 1)}{2\sigma_y^2} e^{2(\sigma_y^2 - \mu_y)} - e^{\sigma_y^2 - 2\mu_y} \right] \tag{7-86}$$

$$\text{var } \sigma_y^2 = \frac{\sigma_y^2}{nD} \left[ (\sigma_y^2 + 1)\, e^{2(\sigma_y^2 - \mu_y)} - e^{\sigma_y^2 - 2\mu_y} \right] \tag{7-87}$$

$$\text{cov }(a, \mu_y) = - \frac{e^{\sigma_y^2/2 - \mu_y}}{2nD} \tag{7-88}$$

$$\text{cov }(a, \sigma_y^2) = \frac{\sigma_y^2}{nD}\, e^{\sigma_y^2/2 - \mu_y} \tag{7-89}$$

$$\mathrm{cov}\,(\mu_y, \sigma_y^2) = -\frac{\sigma_y^2}{nD} e^{\sigma_y^2 - 2\mu_y} \qquad (7\text{-}90)$$

Substitution into 7-62 then yields the variance of the T-year event.

Table 7-2 shows that, as would be expected, the standard error of estimate by method of moments is larger than by maximum likelihood.

## References


1. Burges, S. J., Lettenmaier, D. P., and C. L. Bates, 1975, Properties of the Three-Parameter Lognormal Distribution, Wat. Res. Res., Vol. 11, No. 2, pp. 229-235.

2. Condie, R., 1973, Unpublished Notes, Environment Canada, Ottawa.

3. Sangal, B. P. and A. K. Biswas, 1970, The 3-Parameter Lognormal Distribution and its Applications in Hydrology, Wat. Res. Res., Vol. 6, No. 2, pp. 505-515.

```
      PROGRAM LN3(INPUT,OUTPUT,TAPE5=INPUT,TAPE6=OUTPUT)
C
C
C     COMPUTES METHOD OF MOMENTS AND MAXIMUM LIKELIHOOD ESTIMATES FOR
C     T YEAR EVENTS AND STANDARD ERRORS FOR 3 PARAMETER LOGNORMAL
C     DISTRIBUTION
C     INPUT
C     TITLE
C     N NUMBER OF ANNUAL MAXIMUM EVENTS
C     X SERIES OF EVENTS
C
C
      REAL M1,M2,M3,M4,M5,M6,MY,K,MU
      DIMENSION SND(6), X(100)
      DIMENSION XT(6), SX(6), TITLE(80)
      SND(1)=0.0
      SND(2)=0.8416
      SND(3)=1.2816
      SND(4)=1.6449
      SND(5)=2.0538
      SND(6)=2.3264
      READ (5,16) TITLE
      READ (5,17) N
      XN=N
      READ (5,18) (X(I),I=1,N)
      WRITE (6,19) TITLE
      WRITE (6,20)
      A=0.0
      B=0.0
      C=0.0
      DO 1 I=1,N
      A=A+X(I)
      B=B+X(I)**2
      C=C+X(I)**3
1     CONTINUE
      M1=A/XN
      M2=(B/XN)-(A/XN)**2
      M3=(C/XN)+2.0*M1**3-3.0*M1*(B/XN)
      G=M3/(M2**1.5)
      M2=M2*XN/(XN-1.0)
      WRITE (6,11) M1
      WRITE (6,12) M2
      WRITE (6,13) G
      IF (G.LT.0.0) GO TO 3
      W=(-G+((G**2)+4.0)**0.5)/2.0
      Z2=(1.0-W**(2./3.))/(W**(1./3.))
      AMO=M1-(M2**0.5)/Z2
      WRITE (6,21) AMO
      SY=(ALOG(Z2**2+1.0))**0.5
      SY2=SY**2
      MY=ALOG((M2**0.5)/Z2)-0.5*ALOG(Z2**2+1.0)
      E=EXP(SY2)
      EA=EXP(2.0*SY2)
      EB=EXP(2.5*SY2)
      EC=EXP(3.0*SY2)
      ED=EXP(4.0*SY2)
      EF=EXP(6.0*SY2)
      EG=EXP(10.0*SY2)
      EH=EXP(15.0*SY2)
      EI=EXP(4.0*MY)
      EJ=EXP(5.0*MY)
      EK=EXP(6.0*MY)
      EL=(EXP(SY2)-1.0)**2
      M4=EA*EI*EL*(ED+2.0*EC+3.0*EA-3.0)
      M5=EB*EJ*(EG-5.0*EF+10.0*EC-10.0*E+4.0)
      M6=EC*EK*(EH-6.0*EG+15.0*EF-20.0*EC+15.0*E-5.0)
      VM1=M2/XN
      VM2=(M4-M2**2)/XN
      VM3=(M6-M3**2-6.0*M4*M2+9.0*M2**3)/XN
      CM1M2=M3/XN
      CM1M3=(M4-3.0*M2**2)/XN
      CM2M3=(M5-4.0*M3*M2)/XN
      DO 2 J=1,6
      T=SND(J)
      DXDM1=1.0
      DWDG=-0.5+G/(2.0*(G**2+4.0)**0.5)
      DZ2DW=(-1./3.)*(W**(-4./3.)+W**(-2./3.))
      D1=ALOG(Z2**2+1.0)
      D2=EXP((SQRT(D1))*T-D1/2)
      D3=(2.0*Z2)/(1.0+Z2**2)
```

```
      D4=T/(2.0*Z2*SQRT(D1))
      D5=1.0/Z2
      D6=1.0/(2.0*Z2**3)
      DKDZ2=D3*(D2*(D4-D5-D6)+D6+D5/2.)
      K=(D2-1.0)/Z2
      DKDG=DKDZ2*DZ2DW*DWDG
      DXDM2=(1.0/(2.0*SQRT(M2)))*(K-3.0*G*DKDG)
      DXDM3=DKDG/M2
      SX(J)=SQRT((DXDM1**2)*VM1+(DXDM2**2)*VM2+(DXDM3**2)*VM3+2.0*DXDM1*
     1DXDM2*CM1M2+2.0*DXDM1*DXDM3*CM1M3+2.0*DXDM2*DXDM3*CM2M3)
    2 XT(J)=M1+K*M2**0.5
      WRITE (6,24)
      WRITE (6,25) (XT(J),J=1,6)
      WRITE (6,26) (SX(J),J=1,6)
      GO TO 4
    3 WRITE (6,14)
    4 WRITE (6,27)
      WRITE (6,28)
      XMIN=10000000.
      DO 5 I=1,N
    5 IF (X(I).LT.XMIN) XMIN=X(I)
      AML=XMIN*0.80
      ICOUNT=0
    6 ICOUNT=ICOUNT+1
      A=0.0
      B=0.0
      C=0.0
      D=0.0
      E=0.0
      F=0.0
      P=0.0
      DO 7 I=1,N
      A=A+ALOG(X(I)-AML)
      B=B+(ALOG(X(I)-AML))**2
      P=P+(ALOG(X(I)-AML))**3
      C=C+1.0/((X(I)-AML))
      D=D+1.0/((X(I)-AML)**2)
      E=E+(1.0/((X(I)-AML)))*ALOG(X(I)-AML)
    7 F=F+(1.0/((X(I)-AML)**2))*ALOG(X(I)-AML)
      G=(B/XN)-(A/XN)**2-(A/XN)
      H=(-2.0*E/XN)+(2.0*A/XN)*(C/XN)+(C/XN)
      FCN=C*G+E
      FPN=C*H+D*G+F-D
      AS=AML-(FCN/FPN)
      WRITE (6,29) ICOUNT,AS,FCN
      DELTA=ABS(0.00001*AS)
      IF (ABS(AS-AML).LT.DELTA) GO TO 8
      IF (ICOUNT.GT.25) GO TO 10
      AML=AS
      GO TO 6
    8 CONTINUE
      AML=AS
      MU=A/XN
      VAR=(B/XN)-(A/XN)**2
      VAR=VAR*XN/(XN-1)
      SKEW=(P/XN)+2.0*MU**3-3.0*MU*(B/XN)
      SD=SQRT(VAR)
      A=EXP(VAR-2.0*MU)
      B=EXP(2.0*VAR-2.0*MU)
      C=EXP(VAR/2.0-MU)
      D1=(VAR+1.0)/(2.0*VAR)
      D2=1.0/(2.0*VAR)
      D=D1*B-D2*A-A
      E=1.0/(N*D)
      VA=E*0.5
      VMU=(VAR*E)*(D1*B-A)
      VVAR=VAR*E*((VAR+1.0)*B-A)
      CAMU=C*E/2.0
      CAVAR=VAR*E*C
      CMUVAR=VAR*E*A
      CAMU=-CAMU
      CMUVAR=-CMUVAR
      DO 9 J=1,6
      T=SND(J)
      Z=EXP(MU+T*SD)
      VX=VA+(VMU*Z**2)+(T*Z*CAVAR/SD)+(2.0*Z*CAMU)+(T*Z**2*CMUVAR/SD)+(T
     1**2*Z**2*VVAR/(4.0*VAR))
      XT(J)=AML+Z
      SX(J)=SQRT(VX)
    9 CONTINUE
```

```
      WRITE (6,30)
      WRITE (6,32) AML
      WRITE (6,22) MU
      WRITE (6,23) VAR
      WRITE (6,15) SKEW
      WRITE(6,33)
      WRITE (6,24)
      WRITE (6,25) (XT(J),J=1,6)
      WRITE (6,26) (SX(J),J=1,6)
      WRITE (6,31)
10    CONTINUE
      STOP
C
C
C
11    FORMAT (20X,9HMEAN OF X,16X,E12.5)
12    FORMAT (20X,13HVARIANCE OF X,12X,E12.5)
13    FORMAT (20X,9HSKEW OF X,16X,E12.5)
14    FORMAT (/,3X,52H NO MOMENTS SOLUTION IS POSSIBLE BECAUSE OF -VE SK
     1EW,/)
15    FORMAT (20X,15HSKEW OF LN(X-A),10X,E12.5)
16    FORMAT (80A1)
17    FORMAT (I5)
18    FORMAT (8F10.0)
19    FORMAT (1H1,/,80A1,//,21X,38HTHREE PARAMETER LOGNORMAL DISTRIBUTIO
     1N,/)
20    FORMAT (31X,17HMETHOD OF MOMENTS,//)
21    FORMAT (20X,1HA,24X,E12.5,/)
22    FORMAT (20X,15HMEAN OF LN(X-A),10X,E12.5)
23    FORMAT (20X,19HVARIANCE OF LN(X-A),6X,E12.5)
24    FORMAT (3X,7HT,YEARS,4X,1H2,11X,1H5,10X,2H10,10X,2H20,10X,2H50,9X,
     13H100,/)
25    FORMAT (3X,1HX,3X,6E12.5,/,4X,1HT)
26    FORMAT (3X,1HS,3X,6E12.5,/,4X,1HT,//)
27    FORMAT (25X,28HMAXIMUM LIKELIHOOD PROCEDURE,//)
28    FORMAT (21X,5HTRIAL,11X,1HA,11X,4HF(A),/)
29    FORMAT (22X,I2,8X,E12.5,1X,E12.5)
30    FORMAT (//)
31    FORMAT (/1H1)
32    FORMAT (20X,1HA,24X,E12.5)
33    FORMAT(/,3X,73HFOR GOOD USE OF THIS DISTRIBUTION SKEW OF LN(X-A) S
     1HOULD BE CLOSE TO ZERO,/)
      END
```

ST. MARYS RIVER AT STILLWATER - STATION NO. 01E0001

THREE PARAMETER LOGNORMAL DISTRIBUTION

METHOD OF MOMENTS

| | |
|---|---|
| MEAN OF X | .14555E+05 |
| VARIANCE OF X | .27320E+08 |
| SKEW OF X | .13216E+01 |
| A | .20039E+04 |

| T,YEARS | 2 | 5 | 10 | 20 | 50 | 100 |
|---|---|---|---|---|---|---|
| $X_T$ | .13590E+05 | .18226E+05 | .21347E+05 | .24372E+05 | .28346E+05 | .31380E+05 |
| $S_T$ | .79167E+03 | .10501E+04 | .13839E+04 | .20814E+04 | .35010E+04 | .48840E+04 |

MAXIMUM LIKELIHOOD PROCEDURE

| TRIAL | A | F(A) |
|---|---|---|
| 1 | .47220E+04 | -.64188E-03 |
| 2 | .39798E+04 | -.25684E-03 |
| 3 | .32367E+04 | -.95979E-04 |
| 4 | .26544E+04 | -.31953E-04 |
| 5 | .23716E+04 | -.82744E-05 |
| 6 | .23193E+04 | -.11311E-05 |
| 7 | .23178E+04 | -.30995E-07 |
| 8 | .23178E+04 | -.25063E-10 |

| | |
|---|---|
| A | .23178E+04 |
| MEAN OF LN(X-A) | .93296E+01 |
| VARIANCE OF LN(X-A) | .16826E+00 |
| SKEW OF LN(X-A) | -.86879E-03 |

FOR GOOD USE OF THIS DISTRIBUTION SKEW OF LN(X-A) SHOULD BE CLOSE TO ZERO

| T,YEARS | 2 | 5 | 10 | 20 | 50 | 100 |
|---|---|---|---|---|---|---|
| $X_T$ | .13585E+05 | .18230E+05 | .21377E+05 | .24440E+05 | .28480E+05 | .31575E+05 |
| $S_T$ | .63762E+03 | .98067E+03 | .14006E+04 | .19599E+04 | .28893E+04 | .37251E+04 |

## CHAPTER 8

## TYPE I EXTREMAL DISTRIBUTION

### Introduction

Suppose that from N samples each containing n events the maximum or minimum event in each sample is selected. As n increases, the distribution of the N maxima or minima approaches a limiting or assymptotic form. The type of the limiting form depends on the type of the initial distribution of the Nn values. The distribution of the maxima or minima is given by the functional equation

$$p^n(x) = P(a_n x + b_n) \qquad (8\text{-}1)$$

where $a_n$ and $b_n$ are functions of n.

Fisher and Tippett (quoted in (15)) have shown that there are three possible solutions to the functional equation. These are known, logically enough, as types I, II and III extremal distributions. The type I distribution is unbounded, the type II has a lower limit and the type III has an upper limit.

The type I distribution (or Gumbel (3)) is often used for maximum type events and results from any initial unlimited distribution of exponential type which converges to an exponential function. Examples of this type of distribution include the normal and lognormal distributions. The derivation of the type I distribution for a simple exponential function can be described (13) as follows:

(a) Let $\varepsilon_1, \varepsilon_2, \ldots, \varepsilon_n$ be a series of independent random variables with cumulative probability distribution given by:

$$P(x) = P(\varepsilon_v \leq y) \qquad (8\text{-}2)$$

(b) Define $X_n$ as the maximum value of $\varepsilon$ in a sample of length n i.e. $\max\limits_{1 \leq v \leq n} \varepsilon_v$ so that

$$P(X_n \leq y) = P(\varepsilon_1 \leq y, \varepsilon_2 \leq y, \ldots, \varepsilon_n \leq y) \qquad (8\text{-}3)$$

or

$$P(X_n \leq y) = [P(y)]^n \qquad (8\text{-}4)$$

(c) Now assume that the tail of the distribution P(y) is exponential such that

$$P(y) = 1 - \alpha e^{-y} \qquad (8\text{-}5)$$

(d) From Equation 8-4 if ln (αn) is a normalizing constant

$$P(X_n \leq y + \ln(\alpha n)) = [P(y + \ln(\alpha n))]^n \qquad (8\text{-}6)$$

and from Equation 8-5

$$P(y + \ln(\alpha n)) = 1 - \alpha e^{-(y + \ln(\alpha n))} \qquad (8\text{-}7)$$

so that

$$P(X_n \leq y + \ln(\alpha n)) = [1 - \alpha e^{-(y + \ln(\alpha n))}]^n \qquad (8\text{-}8)$$

or

$$P(X_n \leq y + \ln(\alpha n)) = [1 - e^{-y}/n]^n \qquad (8\text{-}9)$$

(e) If $n \to \infty$, then:

$$\lim_{n\to\infty} P(X_n \leq y + \ln(\alpha n)) = \lim_{n\to\infty} [1 - e^{-y}/n]^n \qquad (8\text{-}10)$$

or

$$\lim_{n\to\infty} P(X_n \leq y + \ln(\alpha n)) = e^{-e^{-y}} \qquad (8\text{-}11)$$

This is the reduced from of the cumulative probability distribution. Substituting the expression

$$y = \alpha(x-\beta) \qquad (8\text{-}12)$$

where α is a concentration parameter and β is a measure of central tendency, the cumulative probability of the type I extremal distribution becomes

$$P(x) = e^{-e^{-\alpha(x-\beta)}} \qquad (8\text{-}13)$$

and the probability density becomes

$$p(x) = \alpha e^{\{-\alpha(x-\beta)-e^{-\alpha(x-\beta)}\}} \qquad (8\text{-}14)$$

## Estimation of Parameters

*Method of Moments* - Applying the standard equation for generation of origin moments to the pdf of the reduced type I extremal distribution gives

$$\mu'_{y,r} = \int_{-\infty}^{\infty} y^r \, e^{-y-e^{-y}} \, dy \qquad (8\text{-}15)$$

Substituting z for $e^{-y}$ results in

$$\mu'_{y,r} = \int_0^{\infty} (-\ln z)^r \cdot z \cdot e^{-z} \cdot -\frac{1}{z} \, dz \qquad (8\text{-}16)$$

The first moment about the origin is then

$$\mu'_{y,1} = \int_0^{\infty} \ln z \, . \, z \, . \, e^{-z} \, . \, \frac{dz}{z} \qquad (8\text{-}17)$$

but $\int_0^{\infty} z\, e^{-z}\, dz$ is $\Gamma(1)$ so that, from Kendall and Stuart (6),

$$\mu'_{y,1} = -\psi(1) = \gamma_E \qquad (8\text{-}18)$$

where $\gamma_E$ is Eulers constant, approximately 0.5772157. Reconverting to the original variate x as $x = y/\alpha + \beta$

$$\mu'_1 = \beta + \gamma_E/\alpha \qquad (8\text{-}19)$$

Similarly, Gumbel (3) has shown that the second moment about the mean, $\mu_2$, is given by

$$\mu_2 = \pi^2/(6\alpha^2) \qquad (8\text{-}20)$$

Rearranging gives expressions for the parameters $\alpha$ and $\beta$

$$\alpha = 1.2825/\sigma \qquad (8\text{-}21)$$

and

$$\beta = \mu - 0.4500\,\sigma \qquad (8\text{-}22)$$

where $\mu$ and $\sigma$ are the mean and standard deviation from the sample.

The coefficients of skew and kurtosis for the type I extremal distribution are constants at approximately 1.14 and 5.4 respectively.

*Maximum Likelihood* - The maximum likelihood solution of the type I extremal distribution was first proposed by Kimball (7) in 1946 (quoted in Gumbel (3)) but was not practical until the advent of computers.

The maximum likelihood method of estimating $\alpha$ and $\beta$ postulates that $\alpha$ and $\beta$ should be such that the probability of n individual maximum events $x_1,\ldots,x_n$ actually being observed as n annual peaks should be a maximum. The probability that $x_1$ occurs as an annual peak event is:

$$p(x_1) = \alpha e^{\{-\alpha(x_1-\beta) - e^{-\alpha(x_1-\beta)}} \qquad (8\text{-}23)$$

and for $x_2$:

$$p(x_2) = \alpha e^{\{-\alpha(x_2-\beta) - e^{-\alpha(x_2-\beta)}} \qquad (8\text{-}24)$$

now

$$p(x_1,\ldots,x_n) = p(x_1)\, p(x_2),\ldots,p(x_n) \tag{8-25}$$

so that:

$$P(x_1,\ldots,x_n) = \alpha^n e^{\{-\alpha \sum_{i=1}^{n} (x_i-\beta) - \sum_{i=1}^{n} e^{-\alpha(x_i-\beta)}\}} \tag{8-26}$$

The method of maximum likelihood then takes the logarithm of Equation 8-26, partially differentiates with respect to $\alpha$ and $\beta$ and equates to zero:

$$\ln L = n \ln \alpha - \alpha \sum_{i=1}^{n} (x_i-\beta) - \sum_{i=1}^{n} e^{-\alpha(x_i-\beta)} \tag{8-27}$$

$$\frac{\partial \ln L}{\partial \alpha} = \frac{n}{\alpha} - \sum_{i=1}^{n} (x_i-\beta) + \sum_{i=1}^{n} (x_i-\beta)\, e^{-\alpha(x_i-\beta)} \tag{8-28}$$

$$\frac{\partial \ln L}{\partial \beta} = n\alpha - \alpha \sum_{i=1}^{n} e^{-\alpha(x_i-\beta)} \tag{8-29}$$

Setting Equation 8-29 equal to zero:

$$\sum_{i=1}^{n} e^{-\alpha(x_i-\beta)} = n \tag{8-30}$$

so that:

$$e^{\alpha\beta} = n / \sum_{i=1}^{n} e^{-\alpha x_i} \tag{8-31}$$

or

$$\beta = \frac{1}{\alpha} \ln \left[n / \sum_{i=1}^{n} e^{-\alpha x_i}\right] \tag{8-32}$$

If the arithmetic mean of the series $x_1,\ldots,x_n$ is denoted by $\mu$, then Equation 8-28 can be written as:

$$\frac{\partial \ln L}{\partial \alpha} = \frac{n}{\alpha} - n(\mu-\beta) + e^{\alpha\beta} \sum_{i=1}^{n} (x_i-\beta)\, e^{-\alpha x_i} \tag{8-33}$$

Substituting for $e^{\alpha\beta}$ from Equation 8-31

$$\frac{\partial \ln L}{\partial \alpha} = \frac{n}{\alpha} - n(\mu-\beta) + \frac{n \sum_{i=1}^{n} (x_i-\beta)\, e^{-\alpha x_i}}{\sum_{i=1}^{n} e^{-\alpha x_i}} \qquad (8\text{-}34)$$

Equating this to zero and simplifying:

$$F(\alpha) = \sum_{i=1}^{n} x_i\, e^{-\alpha x_i} - (\mu-1/\alpha) \sum_{i=1}^{n} e^{-\alpha x_i} = 0 \qquad (8\text{-}35)$$

Equation 8-35 cannot be solved for $\alpha$ analytically and so a Taylor's expansion has been used by Panchang (12).

$$F(\alpha_{j+1}) = F(\alpha_j + h_j) \qquad (8\text{-}36)$$

$$F(\alpha_{j+1}) = F(\alpha_j) + h_j\, F'(\alpha_j) \qquad (8\text{-}37)$$

where $F'(\alpha_j)$ is the first order derivative of $F(\alpha)$ with respect to $\alpha$

$$F'(\alpha) = -\sum_{i=1}^{n} x_i^2\, e^{-\alpha x_i} + (\mu-1/\alpha) \sum_{i=1}^{n} x_i\, e^{-\alpha x_i} - \frac{1}{\alpha^2} \sum_{i=1}^{n} e^{-\alpha x_i} \qquad (8\text{-}38)$$

and $\alpha_j$ and $\alpha_{j+1}$ are successive approximations to $\alpha$. The procedure adopted by Panchang is to estimate $\alpha_1$ from the method of moments (18) (see earlier discussion). By evaluating $F(\alpha_1)$ and $F'(\alpha_1)$ from Equations 8-35 and 8-38 then:

$$h_1 = - F(\alpha_1)/F'(\alpha_1) \qquad (8\text{-}39)$$

and

$$\alpha_2 = \alpha_1 + h_1 \qquad (8\text{-}40)$$

This procedure is repeated until sufficiently small value of $F(\alpha_j)$ is obtained when $\beta$ can be obtained from Equation 8-32. In most cases only 3 Or 4 steps will be required.

Samuelsson (13) has described a similar procedure using different first estimates of $\alpha$ and $\beta$ and suggests that the true parameter values can be estimated to within 1% in only 3 iterations.

Other methods of deriving the maximum likelihood parameter estimates are available (3), (17) but these are generally more complex. Leese (9) has described the modifications to the maximum likelihood equations which are needed to accomodate missing data and historic flood records.

*Other Methods* - A simple approximation has been used by Verma and Advani (15) to estimate the parameters $\alpha$ and $\beta$. If $x_{max}$ is the largest event in a series of n maxima and $x_{min}$ is the smallest event, then the reduced events $y_{max}$ and $y_{min}$ are defined as

$$y_{max} = \alpha(x_{max} - \beta) \quad (8\text{-}41)$$

and

$$y_{min} = \alpha(x_{min} - \beta) \quad (8\text{-}42)$$

By taking the probability of exceedence of the largest event, $x_{max}$, as 1/n and the probability of exceedence of the smallest event, $x_{min}$, as 1/1.01, then the following expressions apply

$$y_{max} = - \ln (- \ln (1-1/n)) \quad (8\text{-}43)$$

and

$$y_{min} = - \ln (- \ln (1-1/1.01)) \quad (8\text{-}44)$$

Evaluation of Equations 8-43 and 8-44 and substitution into Equations 8-41 and 8-42 will, by simultaneous solution, produce rough estimates of $\alpha$ and $\beta$. By using the expansion of ln and neglecting terms above second order, Verma and Advani have produced expressions for $\alpha$ and $\beta$.

Yevjevich (18) has described a process by which estimates of $\alpha$ and $\beta$ can be determined graphically. If, in Equation 8-13, $x = \beta$, then

$$P(\beta) = e^{-1} = 0.368 \quad (8\text{-}45)$$

Plot the event magnitudes, x, versus return period, T = (n + 1)/m, on graph paper with a double exponential scale and fit a straight line through the plotted points. By entering the graph at $T = 1/P(\beta) = 2.717$ the value of $\beta$ is determined. The slope of the best fitting straight line is then equal to $1/\alpha$. This method is very easy to apply but its accuracy is not to be compared with the method of maximum likelihood.

Lowery and Nash (10) have compared various methods of estimating the parameters of the type I extremal distribution. They recognize the greater efficiency of the maximum likelihood technique but recommend moments because of the method's simplicity and lack of bias.

## Frequency Factor

From the cumulative probability distribution (Equation 8-11) the expression relating the reduced variable, y, to return period, T, is

$$y_T = -\ln(-\ln((T-1)/T)) \qquad (8\text{-}46)$$

For convenience, Table 8-1 gives values of the reduced variable $y_T$ for some common return periods.

Table 8-1

Values of the Reduced Variable, $y_T$, of the Type I Extremal Distribution for Some Commonly Used Return Periods, T

| Return Period T | Reduced Variable $y_T$ |
|---|---|
| 2 | 0.3665 |
| 5 | 1.4999 |
| 10 | 2.2504 |
| 20 | 2.9702 |
| 50 | 3.9019 |
| 100 | 4.6001 |

If the n recorded events are placed in order of magnitude so that m = 1 for the largest event and m = n for the smallest event then T = (n+1)/m and Equation 8-46 can be written as

$$y_m = -\ln[-\ln\{(n+1-m)/(n+1)\}] \qquad (8\text{-}47)$$

If the mean, $\mu_y$, and the variance, $\sigma_y^2$, of the series $y_m$, m = 1, 2,...,n, are computed from the reduced sample as:

$$\mu_y = \sum_{m=1}^{n} y_m/n \qquad (8\text{-}48)$$

and

$$\sigma_y^2 = \sum_{m=1}^{n} (y_m-\mu_y)^2/n \qquad (8\text{-}49)$$

and if $\mu$ and $\sigma^2$ are the mean and variance of the recorded events, then the parameters $\alpha$ and $\beta$ can be defined (5) as:

$$\alpha = \sigma_y/\sigma \qquad (8\text{-}50)$$

and

$$\beta = \mu-\mu_y/\alpha \qquad (8\text{-}51)$$

Now, introducing these relationships into the equation for the reduced variate y

$$y_m = \alpha(x-\beta) \qquad (8\text{-}52)$$

and rearranging for x, gives:

$$x = \mu + (y_m-\mu_y)\,\sigma/\sigma_y \qquad (8\text{-}53)$$

Comparing Equation 8-53 with the general frequency equation (e.g. Equation 3-30) it is apparent that for the type I extremal distribution the frequency factor, K, is defined as:

$$K = \frac{y_m - \mu_y}{\sigma_y} \qquad (8\text{-}54)$$

Table 8-2

Mean and Standard Deviation of Order Statistics, m/(n+1), for Various Sample Sizes, n

| Sample Size n | Mean $\mu_y$ | Standard Deviation $\sigma_y$ |
|---|---|---|
| 10 | 0.4952 | 0.9496 |
| 15 | 0.5128 | 1.0206 |
| 20 | 0.5236 | 1.0628 |
| 25 | 0.5309 | 1.0914 |
| 30 | 0.5362 | 1.1124 |
| 35 | 0.5403 | 1.1285 |
| 40 | 0.5436 | 1.1413 |
| 45 | 0.5463 | 1.1518 |
| 50 | 0.5485 | 1.1607 |
| 55 | 0.5504 | 1.1682 |
| 60 | 0.5521 | 1.1747 |
| 65 | 0.5535 | 1.1803 |
| 70 | 0.5548 | 1.1854 |
| 75 | 0.5559 | 1.1898 |
| 80 | 0.5569 | 1.1938 |
| 85 | 0.5578 | 1.1974 |
| 90 | 0.5586 | 1.2007 |
| 95 | 0.5593 | 1.2037 |
| 100 | 0.5600 | 1.2065 |

Since $\mu_y$ and $\sigma_y$ are functions of the sample size only, they can be tabulated. Table 8-2 shows values of $\mu_y$ and $\sigma_y$ for some typical values of n. Alternatively, for a predetermined set of return periods and samples sizes the frequency factor, K, can be tabulated (5) as in Table 8-3.

As an example, for a sample size of 55, Table 8-2 gives values of $\mu_y$ and $\sigma_y$ of 0.5504 and 1.1681 respectively. For a 100 year return period, Table 8-1 or Equation 8-46 gives the reduced variable, $y_m$, as 4.6001 so that, from Equation 8-54, the frequency factor, K, is 3.4670. Alternatively this figure can be found directly from Table 8-3. To determine the 100 year event magnitude it is then only necessary to estimate the population mean, $\mu$ and standard deviation, $\sigma$ from the 55 recorded events and substitute, $\mu$, K and $\sigma$ in the general frequency equation.

Weiss (16) has devised a convenient nomogram for performing graphically the solution of Equation 8-53 given the sample mean and standard deviation. Shown here as Figure 8-1 this nomogram is entered on the left hand side with the required return period, T. From the intersection of the horizontal line through T with the slanting line through the appropriate sample size, n, draw a vertical line to intersect the sloping line corresponding to the sample standard deviation, $\sigma$. From this second intersection draw a horizontal line to cut the right hand edge of the diagram at the value of $K\sigma$, the numerical value to be added to the mean, $\mu$, to give the required event magnitude, x. Examples of the use of this nomogram are given in Weiss (16) and Kendall(5).

Chow (1) has considered the type I extremal distribution as a special case of the lognormal distribution for which the coefficient of skew, $\gamma_1$, is a constant at 1.1396. From the cumulative probability distribution (Equation 8-13) the event magnitude $x_T$ is related to return period, T, by:

$$x_T = \beta - \frac{1}{\alpha} \ln (- \ln (1-1/T)) \qquad (8\text{-}55)$$

where T is 1/[1-P(x)]. Substituting the moments solutions for $\alpha$ and $\beta$ leads to:

$$x_T = \mu - \{0.45 + 0.7797 \ln (- \ln [1-1/T])\} \sigma \qquad (8\text{-}56)$$

Comparison with the standard frequency equation shows that the frequency factor for the type I extremal distribution by this method is:

$$K = - \{0.45 + 0.7797 \ln (- \ln [1-1/T])\} \qquad (8\text{-}57)$$

The last line of Table 8-3 gives the value of the frequency factor determined from this equation. These values correspond to the asymptotic results of using Equation 8-54 as $n \to \infty$.

## Standard Error

*Method of Moments* - The general equation for the standard error of estimate of a 2-parameter distribution has been given earlier as:

$$S_T^2 = \frac{\mu_2}{n} \{1 + K\gamma_1 + \frac{K^2}{4} [\gamma_2-1]\} \qquad (8\text{-}58)$$

Table 8-3

Frequency Factor for Type I Extremal Distribution

| Sample Size n | Cumulative Probability, P, % | | | | | |
|---|---|---|---|---|---|---|
| | 50 | 80 | 90 | 95 | 98 | 99 |
| | Corresponding Return Period, T, Years | | | | | |
| | 2 | 5 | 10 | 20 | 50 | 100 |
| 10 | -0.1355 | 1.0580 | 1.8483 | 2.6063 | 3.5874 | 4.3227 |
| 15 | -0.1434 | 0.9672 | 1.7025 | 2.4078 | 3.3208 | 4.0049 |
| 20 | -0.1478 | 0.9187 | 1.6248 | 2.3020 | 3.1787 | 3.8356 |
| 25 | -0.1506 | 0.8879 | 1.5754 | 2.2350 | 3.0886 | 3.7284 |
| 30 | -0.1526 | 0.8664 | 1.5410 | 2.1881 | 3.0257 | 3.6534 |
| 35 | -0.1540 | 0.8504 | 1.5154 | 2.1532 | 2.9789 | 3.5976 |
| 40 | -0.1552 | 0.8379 | 1.4954 | 2.1261 | 2.9425 | 3.5543 |
| 45 | -0.1561 | 0.8279 | 1.4794 | 2.1044 | 2.9133 | 3.5195 |
| 50 | -0.1568 | 0.8197 | 1.4663 | 2.0865 | 2.8892 | 3.4908 |
| 55 | -0.1574 | 0.8128 | 1.4552 | 2.0714 | 2.8690 | 3.4667 |
| 60 | -0.1580 | 0.8069 | 1.4458 | 2.0586 | 2.8518 | 3.4461 |
| 65 | -0.1584 | 0.8018 | 1.4376 | 2.0475 | 2.8368 | 3.4284 |
| 70 | -0.1588 | 0.7974 | 1.4305 | 2.0377 | 2.8238 | 3.4128 |
| 75 | -0.1592 | 0.7934 | 1.4242 | 2.0291 | 2.8122 | 3.3991 |
| 80 | -0.1595 | 0.7900 | 1.4185 | 2.0215 | 2.8020 | 3.3868 |
| 85 | -0.1597 | 0.7868 | 1.4135 | 2.0146 | 2.7928 | 3.3758 |
| 90 | -0.1600 | 0.7840 | 1.4090 | 2.0084 | 2.7844 | 3.3659 |
| 95 | -0.1602 | 0.7814 | 1.4048 | 2.0028 | 2.7769 | 3.3569 |
| 100 | -0.1604 | 0.7791 | 1.4011 | 1.9977 | 2.7700 | 3.3487 |
| Corresponding Frequency Factors from Equation 8.57 | | | | | | |
| | -0.1643 | 0.7194 | 1.3046 | 1.8658 | 3.5923 | 3.1667 |

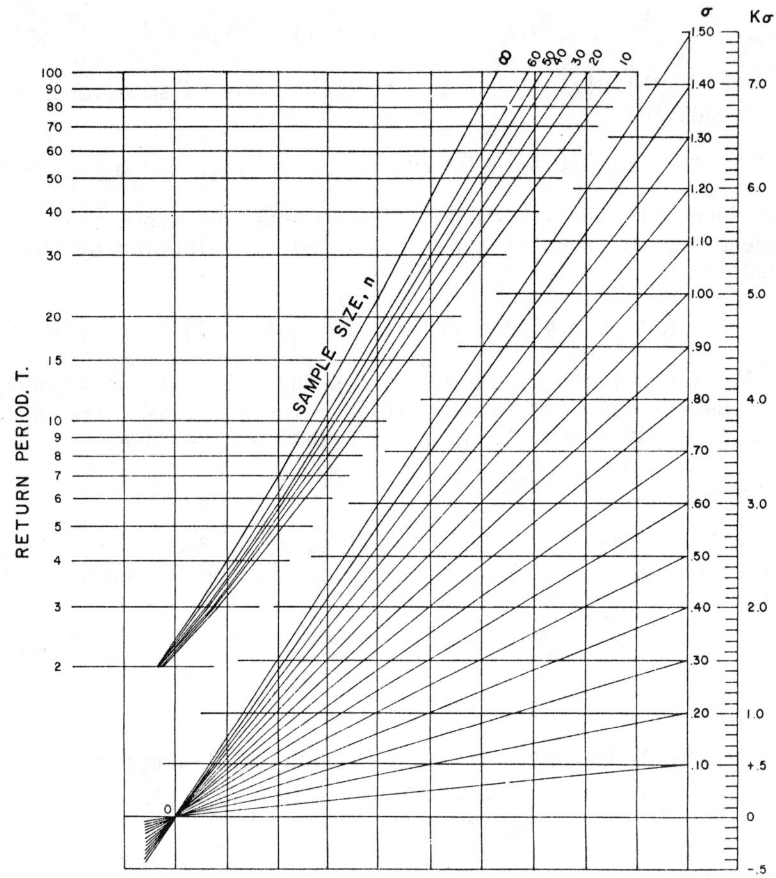


Figure 8-1. Nomogram for use with type I extremal distribution.

From the method of moments,

$$\gamma_1 = 1.1396 \tag{8-59}$$

$$\gamma_2 = \mu_4/\mu_2^2 = 5.4002 \tag{8-60}$$

So that:

$$S_T^2 = \frac{\sigma^2}{n}\,[1 + 1.1396K + 1.1000K^2] \tag{8-61}$$

Taking the square root and simplifying to:

$$S_T = \delta \sqrt{\frac{\sigma^2}{n}} \tag{8-62}$$

values of $\delta$ depend only on K and can thus be predetermined and tabulated for typical values of return period, T, and sample size, n (see Table 8-4). Kaczmarek (4) and Kendall (5) have given similar tables, although as pointed out by Lowery and Nash (10) there is a computational error in the table of Kaczmarek.

Using the same example as before, the 100 year return period event computed from a 55 year sample will have an $\delta$ value of 4.265. Knowing the standard deviation of the recorded events, $\sigma$, the standard error, $S_T$, is computed from Equation 8-62 and the 95% confidence limits are applied as $x_T \pm 1.96\ S_T$.

*Maximum Likelihood* - Since the coefficient of skew is a constant for the type I extremal distribution the maximum likelihood estimate of the standard error of estimate is

$$S_T^2 = \left(\frac{\partial x}{\partial \alpha}\right)^2 \text{var}\ \alpha + \left(\frac{\partial x}{\partial \beta}\right)^2 \text{var}\ \beta + 2\ \frac{\partial x}{\partial \alpha}\frac{\partial x}{\partial \beta}\ \text{cov}\ (\alpha,\beta) \qquad (8\text{-}63)$$

and since from 8-52

Table 8-4

Paramater $\delta$ for Use in Standard Error of Type I Extremal Distribution

| Sample Size n | Cumulative Probability, P, % | | | | | |
|---|---|---|---|---|---|---|
| | 50 | 80 | 90 | 95 | 98 | 99 |
| | Corresponding Return Period, T, Years | | | | | |
| | 2 | 5 | 10 | 20 | 50 | 100 |
| 10 | 0.9305 | 1.8539 | 2.6199 | 3.3826 | 4.3869 | 5.1459 |
| 15 | 0.9269 | 1.7695 | 2.4756 | 3.1814 | 4.1127 | 4.8174 |
| 20 | 0.9250 | 1.7249 | 2.3990 | 3.0745 | 3.9670 | 4.6427 |
| 25 | 0.9238 | 1.6968 | 2.3506 | 3.0069 | 3.8747 | 4.5320 |
| 30 | 0.9229 | 1.6772 | 2.3169 | 2.9597 | 3.8103 | 4.4548 |
| 35 | 0.9223 | 1.6627 | 2.2919 | 2.9247 | 3.7624 | 4.3974 |
| 40 | 0.9218 | 1.6514 | 2.2725 | 2.8975 | 3.7252 | 4.3527 |
| 45 | 0.9214 | 1.6424 | 2.2569 | 2.8756 | 3.6954 | 4.3169 |
| 50 | 0.9211 | 1.6350 | 2.2441 | 2.8577 | 3.6708 | 4.2874 |
| 55 | 0.9208 | 1.6288 | 2.2333 | 2.8426 | 3.6502 | 4.2627 |
| 60 | 0.9206 | 1.6235 | 2.2241 | 2.8297 | 3.6326 | 4.2415 |
| 65 | 0.9204 | 1.6189 | 2.2162 | 2.8186 | 3.6173 | 4.2232 |
| 70 | 0.9202 | 1.6149 | 2.2093 | 2.8089 | 3.6040 | 4.2073 |
| 75 | 0.9201 | 1.6114 | 2.2032 | 2.8003 | 3.5923 | 4.1931 |
| 80 | 0.9199 | 1.6083 | 2.1977 | 2.7926 | 3.5818 | 4.1806 |
| 85 | 0.9198 | 1.6055 | 2.1928 | 2.7858 | 3.5724 | 4.1693 |
| 90 | 0.9197 | 1.6030 | 2.1884 | 2.7796 | 3.5639 | 4.1591 |
| 95 | 0.9196 | 1.6007 | 2.1844 | 2.7739 | 3.5562 | 4.1498 |
| 100 | 0.9195 | 1.5986 | 2.1808 | 2.7688 | 3.5491 | 4.1414 |

$$x_T = \beta + y_T/\alpha \qquad (8\text{-}64)$$

$$\frac{\partial x}{\partial \alpha} = - y_T/\alpha^2 \qquad (8\text{-}65)$$

$$\frac{\partial x}{\partial \beta} = 1 \qquad (8\text{-}66)$$

and

$$S_T^2 = \frac{y_T^2}{\alpha^4} \text{ var } \alpha + \text{var } \beta - 2 \frac{y_T}{\alpha^2} \text{ cov } (\alpha,\beta) \qquad (8\text{-}67)$$

From the maximum likelihood equation (Equation 8-27) the partial derivatives are as follows:

$$\frac{\partial^2 \ln L}{\partial \alpha^2} = - \frac{n}{\alpha^2} - \sum_{i=1}^{n} (x_i - \beta)^2 e^{-\alpha(x_i-\beta)} \qquad (8\text{-}68)$$

$$\frac{\partial^2 \ln L}{\partial \beta^2} = - \alpha^2 \sum_{i=1}^{n} e^{-\alpha(x_i-\beta)} \qquad (8\text{-}69)$$

$$\frac{\partial^2 \ln L}{\partial \alpha \partial \beta} = n + \alpha \sum_{i=1}^{n} (x_i - \beta) e^{-\alpha(x_i-\beta)} \qquad (8\text{-}70)$$

Substituting from Equation 8-30

$$\frac{\partial^2 \ln L}{\partial \beta^2} = - n\alpha^2 \qquad (8\text{-}71)$$

and from Kimball (8)

$$\frac{\partial^2 \ln L}{\partial \alpha^2} = - \frac{1.8237\, n}{\alpha^2} \qquad (8\text{-}72)$$

$$\frac{\partial^2 \ln L}{\partial \alpha \partial \beta} = - .4228\, n \qquad (8\text{-}73)$$

Inverting the likelihood or information matrix leads to the expressions for variances and covariance:

$$\text{var } \alpha = \frac{0.6079\, \alpha^2}{n} \qquad (8\text{-}74)$$

$$\text{var } \beta = \frac{0.2570}{n} \qquad (8\text{-}75)$$

and

$$\text{cov } (\alpha,\beta) = \frac{1.1086}{n\alpha^2} \qquad (8\text{-}76)$$

Substituting these values in Equation 8-67 leads to:

$$S_T^2 = \frac{1}{n\alpha^2} [1.1086 + 0.5140\ y_T + 0.6079\ y_T^2] \qquad (8\text{-}77)$$

which is equivalent to the expression given by Kimball (8):

$$S_T^2 = \frac{1}{n\alpha^2} [1 + (1 - \gamma_E + y_T)^2/(\pi^2/6)] \qquad (8\text{-}78)$$

where $\gamma_E$ is the Euler constant of approximately 0.5772. Simplifying Equations 8-77 and 8-78 to

$$S_T = \delta_\ell / \sqrt{n\alpha^2} \qquad (8\text{-}79)$$

Table 8-5 gives values of $\delta_\ell$ for some values of T. For a comparison with moment estimates, Equation 8-21 was used to replace $\alpha$ with the sample standard deviation. The last line in Table 8-5 shows the resulting values of $\delta_m$ in Equation 8-80.

$$S_T = \delta_m \sqrt{\frac{\sigma^2}{n}} \qquad (8\text{-}80)$$

Table 8-5

| Cumulative Probability, P, % | | | | | |
|---|---|---|---|---|---|
| 50 | 80 | 90 | 95 | 98 | 99 |
| **Corresponding Return Period, T, Years** | | | | | |
| 2 | 5 | 10 | 20 | 50 | 100 |
| **Value of $\delta_\ell$ in Equation 8-79** | | | | | |
| 1.1742 | 1.8020 | 2.3118 | 2.8282 | 3.5171 | 4.0420 |
| **Value of $\delta_m$ in Equation 8-80** | | | | | |
| 0.7139 | 1.0956 | 1.4055 | 1.7195 | 2.1383 | 2.4574 |

Comparison with Table 8-4 shows that the values of $\delta_m$, the moments solution, are larger than $\delta_\ell$, the likelihood solution.

*Other Methods* - Using the relationship

$$\text{cov}\ (\mu,\sigma) = \rho(\text{var}\ \mu)^{\frac{1}{2}}\ (\text{var}\ \sigma)^{\frac{1}{2}} \qquad (8\text{-}81)$$

where $\rho$ is the simple linear correlation coefficient between $\mu$ and $\sigma$, Nash and Amorocho (11) found that $\rho$ has a mean value of 0.56 and a standard deviation of 0.02. Substitution in the standard error relationship (Equation 8-58) yields

$$S_T^2 = \frac{\sigma^2}{n} (1 + 1.18K + 1.1K^2) \qquad (8\text{-}82)$$

which approaches the theoretical relationship of Equation 8-61.

Dalrymple (2) has given the following equation for the standard error of the reduced variate, y, of the type I extremal distribution

$$S' = \frac{e^y}{\sqrt{n}} \sqrt{\frac{1}{T - 1}} \tag{8-83}$$

The confidence interval for the reduced variable is then computed as

$$y_T \pm t\,S' \tag{8-84}$$

Dalrymple used this confidence region to determine the homogeneous hydrologic region in his index flood method of regional flood frequency analysis (see Chapter 12).

References

1. Chow, V. T., 1954, The Log-Probability Law and its Engineering Applications, Proc. ASCE, Vol. 80, pp. 1-25.

2. Dalrymple, T., 1960, Flood Frequency Analysis, USGS Water Supply Paper 1543-A.

3. Gumbel, E. J., 1958, Statistics of Extremes, Columbia University Press.

4. Kaczmarek, Z., 1958, Efficiency of the Estimation of Floods with a Given Return Period, Proc. AISH General Assembly of Toronto, 1957, Vol. III, 144-159.

5. Kendall, G. R., 1959, Statistical Analysis of Extreme Values, Proc. Hydrology Symp. No. 1, Spillway Design Floods, NRC, Ottawa, pp. 54-78.

6. Kendall, M. G. and A. Stuart, 1963, The Advanced Theory of Statistics, Vol. I, Griffin, London.

7. Kimball, B. F., 1946, Sufficient Statistical Estimation Functions for the Parameters of the Distribution of Maximum Values, Ann. Math. Stats., Vol. 17, pp. 299-309.

8. Kimball, B. F., 1949, An Approximation to the Sampling Variances of an Estimated Maximum Value of Given Frequency Based on fit of Doubly Exponential Distribution of Maximum Values, Ann. Math. Stats., Vol. 20, pp. 110-113.

9. Leese, M. N., 1973, The Use of Censored Data in Estimating T-Year Floods, Proceedings of the UNESCO/WMO/IAHS

Symposium on the Design of Water Resources Projects with Inadequate Data, Madrid, Vol. 1, pp. 235-247.

10. Lowery, M. D. and J. E. Nash, 1970, A Comparison of Methods of Fitting the Double Exponential Distribution, Journal of Hydrology, Vol. 10, No. 3, pp. 259-275.

11. Nash, J. E. and J. Amorocho, 1966, TheAccuracy of the Prediction of Floods of High Return Period, Wat. Res. Res., Vol. 2, No. 2, pp. 191-198.

12. Panchang, G. M., 1967, Improved Precision of Future High Floods, Proc. Symp. on Floods and their Computations, Leningrad, pp. 51-59.

13. Samuelsson, B., 1972, Statistical Interpretation of Hydrometeorological Extreme Values, Nordic Hydrology, Vol. 3, No. 4, pp. 199-214.

14. Todorovic, P. and J. Rousselle, 1971, Some Problems of Flood Analysis, Wat. Res. Res., Vol. 7, No. 5, pp. 1144-1150.

15. Verma, R. D. and R. M. Advani, 1973, Flood Control Planning with Inadequate Hydrologic Data, Proceedings of Second International Symposium in Hydrology, Water Resources Publications, Fort Collins, Colorado 80521, pp. 259-268.

16. Weiss, L. L., 1955, A Nomogram Based on the Theory of Extreme Values for Determining Values for Various Return Periods, Monthly Weather Review, Vol. 83, No. 3, pp. 69-71.

17. WMO, 1969, Estimation of Maximum Floods, Technical Note No. 98, WMO No. 233.TP.126, Geneva.

18. Yevjevich, V., 1972, Probability and Statistics in Hydrology, Water Resources Publications, Fort Collins, Colorado.

```
      PROGRAM TIE(INPUT,OUTPUT,TAPE5=INPUT,TAPE6=OUTPUT)
C
C
C     COMPUTES METHOD OF MOMENTS AND MAXIMUM LIKELIHOOD ESTIMATES FOR
C     T YEAR EVENTS AND STANDARD ERRORS FOR TYPE 1 EXTREMAL
C     DISTRIBUTION
C     INPUT
C     TITLE
C     N NUMBER OF ANNUAL MAXIMUM EVENTS
C     X SERIES OF EVENTS
C
C
      REAL M1,M2,K
      DIMENSION T(6), X(100)
      DIMENSION XT(6), SX(6), TITLE(80)
      T(1)=2.
      T(2)=5.
      T(3)=10.
      T(4)=20.
      T(5)=50.
      T(6)=100.
      READ (5,9) TITLE
      READ (5,10) N
      XN=N
      READ (5,11) (X(I),I=1,N)
      A=0.0
      B=0.0
      C=0.0
      DO 1 I=1,N
      A=A+X(I)
      B=B+X(I)**2
      C=C+X(I)**3
1     CONTINUE
      M1=A/XN
      M2=(B/XN)-(A/XN)**2
      M2=M2*XN/(XN-1.0)
      M3=(C/XN)+2.0*M1**3-3.0*M1*(B/XN)
      SKEW=M3/(M2**1.5)
      ALPHA=1.2825/(SQRT(M2))
      BETA=M1-0.45*SQRT(M2)
      A=0.0
      B=0.0
      DO 2 I=1,N
      XI=I
      XN=N
      Y=-ALOG(-ALOG((XN+1.0-XI)/(XN+1.0)))
      A=A+Y
      B=B+Y**2
2     CONTINUE
      YBAR=A/XN
      YSTD=SQRT((B/XN)-YBAR**2)
      DO 3 J=1,6
      YM=-ALOG(-ALOG((T(J)-1.0)/T(J)))
      K=(YM-YBAR)/YSTD
      XT(J)=M1+K*SQRT(M2)
      DELTA=1.0+1.139547093*K+1.100000027*K**2
      SX(J)=SQRT(M2*DELTA/XN)
3     CONTINUE
      WRITE (6,12) TITLE
      WRITE (6,13)
      WRITE (6,20) ALPHA,M1
      WRITE (6,21) BETA,M2
      WRITE (6,22) SKEW
      WRITE (6,25)
      WRITE (6,14)
      WRITE (6,15) (XT(J),J=1,6)
      WRITE (6,16) (SX(J),J=1,6)
      WRITE (6,17)
      WRITE (6,18)
      ICOUNT=0
      AML=ALPHA
4     ICOUNT=ICOUNT+1
      A=1.0/(AML**2)
      B=M1-1.0/AML
      C=0.0
      D=0.0
      E=0.0
      DO 5 I=1,N
      TEMP=EXP(-AML*X(I))
      C=C+TEMP
```

```
      D=D+TEMP*X(I)
      E=E+TEMP*X(I)**2
5     CONTINUE
      FCN=D-B*C
      FPN=B*D-E-A*C
      AS=AML-(FCN/FPN)
      WRITE (6,19) ICOUNT,AS,FCN
      DELTA=ABS(0.0000001*AS)
      IF (ABS(AS-AML).LT.DELTA) GO TO 6
      IF (ICOUNT.GT.25) GO TO 8
      AML=AS
      GO TO 4
6     CONTINUE
      ALPHA=AS
      BETA=(1.0/ALPHA)*ALOG(XN/C)
      M2=1.2825/ALPHA
      M1=BETA+0.45*M2
      M2=M2**2
      DO 7 J=1,6
      YM=-ALOG(-ALOG(1.0-1.0/T(J)))
      XT(J)=BETA+YM/ALPHA
      SX(J)=SQRT((1.1086+0.5140*YM+0.6079*YM**2)/(XN*ALPHA**2))
7     CONTINUE
      WRITE (6,23)
      WRITE (6,20) ALPHA,M1
      WRITE (6,24) BETA,M2
      WRITE (6,14)
      WRITE (6,15) (XT(J),J=1,6)
      WRITE (6,16) (SX(J),J=1,6)
8     CONTINUE
      STOP
C
9     FORMAT (80A1)
10    FORMAT (I5)
11    FORMAT (8F10.0)
12    FORMAT (1H1,/,80A1,//,26X,28HTYPE I EXTREMAL DISTRIBUTION,/)
13    FORMAT (31X,17HMETHOD OF MOMENTS,//)
14    FORMAT (3X,7HT,YEARS,4X,1H2,11X,1H5,10X,2H10,10X,2H20,10X,2H50,9X,
     13H100,/)
15    FORMAT (3X,1HX,3X,6E12.5,/,4X,1HT)
16    FORMAT (3X,1HS,3X,6E12.5,/,4X,1HT,//)
17    FORMAT (25X,28HMAXIMUM LIKELIHOOD PROCEDURE,//)
18    FORMAT (21X,5HTRIAL,11X,1HA,11X,4HF(A),/)
19    FORMAT (22X,I2,8X,E12.5,1X,E12.5)
20    FORMAT (9X,5HALPHA,5X,E12.5,14X,4HM1   ,6X,E12.5)
21    FORMAT (9X,5HBETA ,5X,E12.5,14X,4HM2   ,6X,E12.5)
22    FORMAT (45X,4HSKEW,6X,E12.5,/)
23    FORMAT (//)
24    FORMAT (9X,5HBETA ,5X,E12.5,14X,4HM2   ,6X,E12.5,/)
25    FORMAT (3X,67HNOTE - FOR GOOD USE OF THIS DISTRIBUTION SKEW SHOULD
     1 BE AROUND 1.13,/)
      END
```

ST. MARYS RIVER AT STILLWATER - STATION NO. 01E0001

TYPE 1 EXTREMAL DISTRIBUTION

METHOD OF MOMENTS

| | | | |
|---|---|---|---|
| ALPHA | .24537E-03 | M1 | .14555E+05 |
| BETA | .12203E+05 | M2 | .27320E+08 |
| | | SKEW | .12887E+01 |

NOTE - FOR GOOD USE OF THIS DISTRIBUTION SKEW SHOULD BE AROUND 1.13

| T,YEARS | 2 | 5 | 10 | 20 | 50 | 100 |
|---|---|---|---|---|---|---|
| $X_T$ | .13729E+05 | .18772E+05 | .22111E+05 | .25315E+05 | .29460E+05 | .32567E+05 |
| $S_T$ | .62118E+03 | .10955E+04 | .15008E+04 | .19095E+04 | .24512E+04 | .28621E+04 |

MAXIMUM LIKELIHOOD PROCEDURE

| TRIAL | A | F(A) |
|---|---|---|
| 1 | .25583E-03 | .89135E+03 |
| 2 | .25757E-03 | .11216E+03 |
| 3 | .25761E-03 | .26002E+01 |
| 4 | .25761E-03 | .14954E-02 |

| | | | |
|---|---|---|---|
| ALPHA | .25761E-03 | M1 | .14512E+05 |
| BETA | .12271E+05 | M2 | .24785E+08 |

| T,YEARS | 2 | 5 | 10 | 20 | 50 | 100 |
|---|---|---|---|---|---|---|
| $X_T$ | .13694E+05 | .18094E+05 | .21007E+05 | .23801E+05 | .27418E+05 | .30128E+05 |
| $S_T$ | .58842E+03 | .90307E+03 | .11585E+04 | .14173E+04 | .17625E+04 | .20256E+04 |

## CHAPTER 9

## PEARSON TYPE III DISTRIBUTION

### Introduction

The probability density distribution of the Pearson type III distribution is of the form

$$p(x) = \frac{1}{\alpha\Gamma(\beta)} \left\{\frac{x-\gamma}{\alpha}\right\}^{\beta-1} e^{-\left\{\frac{x-\gamma}{\alpha}\right\}} \qquad (9\text{-}1)$$

where $\alpha$, $\beta$ and $\gamma$ are parameters to be defined and $\Gamma(\beta)$ is the gamma function.

If the substitution $y = (x-\gamma)/\alpha$ is made, then Equation 9-1 simplifies to

$$p(y) = \frac{y^{\beta-1}\, e^{-y}}{\Gamma(\beta)} \qquad (9\text{-}2)$$

which is a one parameter gamma distribution described in many statistics texts.

### Estimation of Parameters

*Method of Moments* - Substituting the probability density function of the Pearson type III distribution (Equation 9-1) into the general equation for moments about the origin yields

$$\mu_r' = \int_o^\infty x^r \frac{1}{\alpha\Gamma(\beta)} \left(\frac{x-\gamma}{\alpha}\right)^{\beta-1} e^{-\left(\frac{x-\gamma}{\alpha}\right)} dx \qquad (9\text{-}3)$$

Substituting y for $(x-\gamma)/\alpha$ gives

$$\mu_r' = \frac{1}{\Gamma(\beta)} \int_o^\infty (\alpha y+\gamma)^r\, y^{\beta-1}\, e^{-y}\, dy \qquad (9\text{-}4)$$

which can be evaluated by noting that

$$\int_o^\infty y^{\beta-1}\, e^{-y}\, dy = \Gamma(\beta) \qquad (9\text{-}5)$$

the gamma function. As an example, for r=1, the first moment about the origin, $\mu_1'$, is given by

$$\mu_1' = \frac{1}{\Gamma(\beta)} \int_o^\infty (\alpha y^\beta\, e^{-y} + \gamma y^{\beta-1}\, e^{-y})\, dy \qquad (9\text{-}6)$$

$$\mu_1' = \frac{\alpha\Gamma(\beta+1)}{\Gamma(\beta)} + \gamma\,\frac{\Gamma(\beta)}{\Gamma(\beta)} = \alpha\beta + \gamma \qquad (9\text{-}7)$$

For r = 2,

$$\mu_2' = \alpha^2\beta(\beta+1) + 2\alpha\beta\gamma + \gamma^2 \tag{9-8}$$

and since

$$\mu_2 = \mu_2' - \mu_1'^2 \tag{9-9}$$

the second central moment is given by

$$\mu_2 = \sigma^2 = \alpha^2\beta \tag{9-10}$$

Similarly, higher order central moments can be calculated as

$$\mu_3 = 2\alpha^3\beta \tag{9-11}$$

$$\mu_4 = 3\alpha^4\beta(\beta+2) \tag{9-12}$$

$$\mu_5 = 4\alpha^5\beta(5\beta+6) \tag{9-13}$$

$$\mu_6 = 5\alpha^6\beta(3\beta^2+26\beta+24) \tag{9-14}$$

If moment ratios $\gamma_1 \rightarrow \gamma_4$ are defined as

$$\gamma_1 = \mu_3/\mu_2^{3/2} \tag{9-15}$$

$$\gamma_2 = \mu_4/\mu_2^2 \tag{9-16}$$

$$\gamma_3 = \mu_5/\mu_2^{5/2} \tag{9-17}$$

$$\gamma_4 = \mu_6/\mu_2^3 \tag{9-18}$$

(note that $\gamma_1$ and $\gamma_2$ are the coefficients of skew and kurtosis respectively), then, for the Pearson type III distribution, $\gamma_2$ - $\gamma_4$ can be defined in terms of $\gamma_1$ as

$$\gamma_2 = 3(1+\gamma_1^2/2) \tag{9-19}$$

$$\gamma_3 = \gamma_1(10+3\gamma_1^2) \tag{9-20}$$

$$\gamma_4 = 5(3+13\gamma_1^2/2 + 3\gamma_1^4/2) \tag{9-21}$$

From the sample mean, $\mu$, standard deviation, $\sigma$, and coefficient of skew, $\gamma_1$, the parameters $\alpha$, $\beta$ and $\gamma$ can be determined. Bobée and Robitaille (4) have shown that the sample coefficient of skew, $\gamma_1$ from Equation 9-15 should be corrected for bias as

$$\gamma_1 = \hat{\gamma}_1 \frac{\sqrt{N(N-1)}}{N-2} \left(1 + \frac{8.5}{N}\right) \tag{9-22}$$

where N is the sample size. Then, from 9-10, 9-11 and 9-15

$$\beta = (2/\gamma_1)^2 \tag{9-23}$$

from 9-7 and 9-10

$$\alpha = \sigma/\sqrt{\beta} \tag{9-24}$$

and

$$\gamma = \mu - \sigma\sqrt{\beta} \tag{9-25}$$

*Maximum Likelihood* - The likelihood function is set up, as usual, as the logarithm of the product of the probability density distributions:

$$\ln L = -n \ln \Gamma(\beta) - \frac{1}{\alpha} \sum_{i=1}^{n} (x_i - \gamma) + (\beta - 1) \sum_{i=1}^{n} \ln (x_i - \gamma) - n\beta \ln \alpha \tag{9-26}$$

Differentiating with respect to $\alpha$, $\beta$ and $\gamma$ and equating to zero gives the following three equations (9):

$$\frac{\partial \ln L}{\partial \alpha} = \frac{1}{\alpha^2} \sum_{i=1}^{n} (x_i - \gamma) - \frac{n\beta}{\alpha} = 0 \tag{9-27}$$

$$\frac{\partial \ln L}{\partial \beta} = -n\, \Gamma'(\beta)/\Gamma(\beta) + \sum_{i=1}^{n} \ln (x_i - \gamma) - n \ln \alpha = 0 \tag{9-28}$$

$$\frac{\partial \ln L}{\partial \gamma} = \frac{n}{\alpha} - (\beta - 1) \sum_{i=1}^{n} (1/(x_i - \gamma)) = 0 \tag{9-29}$$

The maximum likelihood solution then depends on a simultaneous solution of these three equations. The psi or digamma function, $\psi(\beta)$ or $\Gamma'(\beta)/\Gamma(\beta)$ in Equation 9-28 can be calculated using an asymptotic equation (1) of the form

$$\psi(\beta) = \ln (\beta) - \frac{1}{2\beta} - \frac{1}{12\beta^2} + \frac{1}{120\beta^4} - \frac{1}{252\beta^6} \tag{9-30}$$

Condie and Nix (6) found that to perserve accuracy at low values of $\beta$ it was necessary to include a recurrence equation so that

$$\psi(\beta) = \ln (\beta+2) - \frac{1}{2(\beta+2)} - \frac{1}{12(\beta+2)^2} + \frac{1}{120(\beta+2)^4} - \frac{1}{252(\beta+2)^6} - \frac{1}{(\beta+1)} - \frac{1}{\beta} \tag{9-31}$$

Eliminating $\alpha$ from Equations 9-27 and 9-29 yields an expression for $\beta$

$$\beta = 1 \Big/ \left[ 1 - \frac{n^2}{\sum_{i=1}^{n} (x_i - \gamma) \cdot \sum_{i=1}^{n} (1/(x_i - \gamma))} \right] \qquad (9\text{-}32)$$

Substituting for $\beta$ in 9-27 leads to

$$\alpha = \frac{\sum_{i=1}^{n} (x_i - \gamma)}{n} - \frac{n}{\sum_{i=1}^{n} (1/(x_i - \gamma))} \qquad (9\text{-}33)$$

Further substitution of $\alpha$ and $\beta$ from these two expressions into Equation 9-28 yields an equation in $\gamma$ only which can be solved numerically at $\partial \ln L/\partial\beta = 0$. Using an iterative procedure such as Newton's method gives a closure although multiple roots sometimes occur.

If the value of $\gamma$ can be derived from a source other than the data then direct maximum likelihood estimates of $\alpha$ and $\beta$ are available. If the arithemtic mean of the reduced series $x-\gamma$ is given by

$$A = \frac{1}{n} \sum_{i=1}^{n} (x - \gamma) \qquad (9\text{-}34)$$

then Equation 9-27 reduces to

$$\alpha = A/\beta \qquad (9\text{-}35)$$

Now if the geometric mean of the reduced series $x-\gamma$ is given by

$$B = \left[ \prod_{i=1}^{n} (x - \gamma) \right]^{1/n} \qquad (9\text{-}36)$$

and if

$$C = \ln A - \ln B \qquad (9\text{-}37)$$

then substitution of Equations 9-35 and 9-37 into Equation 9-28 and simplifying yields

$$\ln \beta - \Gamma'(\beta)/\Gamma(\beta) = C \qquad (9\text{-}38)$$

Solution of Equation 9-38 gives an estimate of $\beta$ and substitution back into Equation 9-35 will result in an estimate of $\alpha$. Solution of this equation is made easier by the availability of tables of C versus $\beta C$. Thus calculation of C from the reduced data and interpolation from a table gives $\beta C$ from which $\beta$ can be quickly found. Greenwood and Durand (7) have also given the following polynomial approximations to estimate $\beta$ directly from C:

for $0 \leq C \leq 0.5772$

$$\hat{\beta} = (0.5000876 + 0.1648852C - 0.054427C^2)/C \qquad (9\text{-}39)$$

for which the maximum error is 0.0088% and

for $0.5772 \leq C \leq 17.0$

$$\hat{\beta} = \frac{8.898919 + 9.059950C + 0.9775373C^2}{C(17.79728 + 11.968477C + C^2)} \qquad (9\text{-}40)$$

for which the maximum error is 0.00554%.

The maximum likelihood methods described above may not always be applicable. Matalas and Wallis (10) have noted that for very small values of sample skew a solution may not be possible. Similarly if $\beta < 1$ a maximum likelihood solution is not possible. If $\beta$ must be greater than 1 then the sample coefficient of skew cannot exceed 2 (see Equation 9-23). Finally if the sample coefficient of skew is negative the Pearson type III distribution becomes bounded at the upper limit which is not suitable for analysis of maximum events. In general, it is not possible with any automatic procedure to guarantee finding the minimum variance solution each time. Each data set requires a careful investigation of the shape of the function.

## Frequency Factor

The cumulative probability distribution of the Pearson type III can be expressed as:

$$P(x) = \frac{1}{\alpha\Gamma(\beta)} \int_0^{x_o} e^{-\{\frac{x-\gamma}{\alpha}\}} \{\frac{x-\gamma}{\alpha}\}^{\beta-1} dx \qquad (9\text{-}41)$$

If $\alpha$, $\beta$ and $\gamma$ are derived as in the previous section and the probability required for a given return period, T, is $P(x) = 1 - 1/T$ then to define the event magnitude, Equation 9-41 must be solved for $x_o$.

Making the substitution $y = (x-\gamma)/\alpha$ the distribution is given by:

$$P(y) = \frac{1}{\Gamma(\beta)} \int_0^{y_o} y^{\beta-1} e^{-y} dy \qquad (9\text{-}42)$$

but from (1)

$$P(y) = P(\chi^2|v) \qquad (9\text{-}43)$$

where $P(\chi^2|v)$ is the chi-square distribution with $2\beta$ degrees of freedom and $\chi^2 = 2y$. So, from the tabulated value of $\chi^2$ for probability 1 - 1/T and $2\beta$ degrees of freedom the reduced event magnitude, $y_o$, is obtained as:

$$y_o = \chi^2/2 \tag{9-44}$$

and the expected event magnitude is given by:

$$x = \frac{\chi^2_\alpha}{2} + \gamma \tag{9-45}$$

Tables of chi-square distribution are commonly given in statistical texts, and, for convenience, Table 9-1 provides

Table 9-1

Percentage Points of the Chi-Square Distribution[1]

| Degrees of Freedom | Cumulative Probability, $P(\chi^2\|v)$, % | | | | | |
|---|---|---|---|---|---|---|
| | 50 | 75 | 90 | 95 | 97.5 | 99 |
| | Corresponding Return Period, T, Years | | | | | |
| v | 2 | 4 | 10 | 20 | 40 | 100 |
| 1 | 0.46 | 1.32 | 2.71 | 3.84 | 5.02 | 6.63 |
| 2 | 1.39 | 2.77 | 4.61 | 5.99 | 7.38 | 9.21 |
| 3 | 2.37 | 4.11 | 6.25 | 7.81 | 9.35 | 11.34 |
| 4 | 3.36 | 5.39 | 7.78 | 9.49 | 11.14 | 13.28 |
| 5 | 4.35 | 6.63 | 9.24 | 11.07 | 12.83 | 15.09 |
| 6 | 5.35 | 7.84 | 10.64 | 12.59 | 14.45 | 16.81 |
| 7 | 6.35 | 9.04 | 12.02 | 14.07 | 16.01 | 18.48 |
| 8 | 7.34 | 10.22 | 13.36 | 15.51 | 17.53 | 20.09 |
| 9 | 8.34 | 11.39 | 14.68 | 16.92 | 19.02 | 21.67 |
| 10 | 9.34 | 12.55 | 15.99 | 18.31 | 20.48 | 23.21 |
| 11 | 10.34 | 13.70 | 17.28 | 19.68 | 21.92 | 24.73 |
| 12 | 11.34 | 14.84 | 18.55 | 21.03 | 23.34 | 26.32 |
| 13 | 12.34 | 15.98 | 19.81 | 22.36 | 24.74 | 27.69 |
| 14 | 13.34 | 17.11 | 21.06 | 23.68 | 26.12 | 29.14 |
| 15 | 14.34 | 18.25 | 22.31 | 25.00 | 27.49 | 30.58 |
| 16 | 15.34 | 19.37 | 23.54 | 26.30 | 28.85 | 32.00 |
| 17 | 16.34 | 20.48 | 24.77 | 27.59 | 30.19 | 33.41 |
| 18 | 17.34 | 21.61 | 25.99 | 29.87 | 31.53 | 34.81 |
| 19 | 18.34 | 22.72 | 27.20 | 30.14 | 32.85 | 36.19 |
| 20 | 19.34 | 23.83 | 28.41 | 31.41 | 34.17 | 37.57 |
| 21 | 20.34 | 24.93 | 29.62 | 32.67 | 35.48 | 38.93 |
| 22 | 21.34 | 26.04 | 30.81 | 33.92 | 36.78 | 40.29 |
| 23 | 22.34 | 27.14 | 32.01 | 35.17 | 38.08 | 41.64 |
| 24 | 23.34 | 28.24 | 33.20 | 36.42 | 39.36 | 42.98 |
| 25 | 24.34 | 29.34 | 34.38 | 37.65 | 40.65 | 44.31 |
| 26 | 25.34 | 30.43 | 35.56 | 38.89 | 41.92 | 45.64 |
| 27 | 26.34 | 31.53 | 36.74 | 40.11 | 43.19 | 46.96 |
| 28 | 27.34 | 32.62 | 37.92 | 41.34 | 44.46 | 48.28 |
| 29 | 28.34 | 33.71 | 39.09 | 42.56 | 45.72 | 49.59 |
| 30 | 29.34 | 34.80 | 40.26 | 43.77 | 46.98 | 50.89 |
| 40 | 39.34 | 45.62 | 51.81 | 55.76 | 59.34 | 63.69 |
| 50 | 49.33 | 56.33 | 63.17 | 67.50 | 71.42 | 76.17 |
| 60 | 59.33 | 66.98 | 74.40 | 79.08 | 83.30 | 88.38 |
| 70 | 69.33 | 77.57 | 85.53 | 90.53 | 95.02 | 100.42 |
| 80 | 79.33 | 88.13 | 96.58 | 101.88 | 106.63 | 112.33 |
| 90 | 89.33 | 98.65 | 107.56 | 113.14 | 118.14 | 124.12 |
| 100 | 99.33 | 109.14 | 118.49 | 124.34 | 129.56 | 135.81 |

[1] The function, $\chi^2_\alpha$, tabulated is that value of $\chi^2$ with v degrees of freedom beyond which (1-P)% of the distribution lies.

values of $\chi^2$ for some commonly used probabilities and for various degrees of freedom. Note that in Table 9-1 the probabilities are arranged so that larger event magnitudes correspond to larger cumulative probabilities and smaller probabilities of exceedence (larger return periods). This table would be suitable for an analysis of maxima such as flood events. In the study of minima such as drought events, for which this distribution is sometimes used, (c.f. Chin (4)), these probabilities should be reversed.

It has been found (8) that the expression

$$\left\{\left(\frac{\chi^2}{v}\right)^{1/3} + \frac{2}{9v} - 1\right\}\left\{\frac{9v}{2}\right\}^{1/2} \qquad (9\text{-}46)$$

(known as the Wilson-Hilferty approximation) is approximately normally distributed with zero mean and unit variance for $v > 30$. Thus the value of $X^2$ for a particular probability level, P, and number of degrees of freedom, v, can be approximated by substituting the corresponding standard normal deviate, t, in the equation:

$$\chi^2 \simeq v\left\{1 - \frac{2}{9v} + t.\sqrt{\frac{2}{9v}}\right\}^3 \qquad (9\text{-}47)$$

A refinement to the approximation may be made by substituting $(t-h_v)$ for t in Equation 9-47 where

$$h_v = \frac{60}{v}\ h_{60} \qquad (9\text{-}48)$$

and $h_{60}$ is tabulated in Abramowitz and Stegun (1).

Combining Equations 9-45 and 9-47 and substituting the degrees of freedom, $v = 2\beta$,

$$x \approx \alpha\beta\left\{1 - \frac{1}{9\beta} + t\sqrt{\frac{1}{9\beta}}\right\}^3 + \gamma \qquad (9\text{-}49)$$

Thus, knowing, $\alpha$, $\beta$ and $\gamma$ the value of x corresponding to any given probability level can be approximated.

Substituting for $\alpha$ and $\gamma$ from Equations 9-24 and 9-25 in Equation 9-45 an expression:

$$x \simeq \mu + [\frac{\chi^2\gamma_1}{4} - \frac{2}{\gamma_1}]\sigma \qquad (9\text{-}50)$$

is obtained, where $\gamma_1$ is the coefficient of skew of the sample data.

Comparing Equation 9-50 with the general frequency equation it can be seen that, for the Pearson type III distribution, the frequency factor, K, can be approximated by

$$K \approx \frac{\chi^2 \gamma_1}{4} - \frac{2}{\gamma_1} \qquad (9\text{-}51)$$

For computation purposes and for later use in calculating standard errors it is more useful to express K directly in terms of $\gamma_1$. By substituting Equation 9-47 in Equation 9-51 and noting that degrees of freedom, v, is equal to $8/\gamma_1^2$ the expression becomes

$$K \doteq t + (t^2-1)\frac{\gamma_1}{6} + \frac{1}{3}(t^3-6t)\left(\frac{\gamma_1}{6}\right)^2$$

$$- (t^2-1)\left(\frac{\gamma_1}{6}\right)^3 + t\left(\frac{\gamma_1}{6}\right)^4 + \frac{1}{3}\left(\frac{\gamma_1}{6}\right)^5 \qquad (9\text{-}52)$$

From this equation tables of K for different coefficients of skew, $\gamma_1$, and probability, P, can be calculated. Table 9-2 is an example. The Soil Conservation Service, U. S. Dept. of Agriculture (13) has prepared very comprehensive tables of frequence factors for the Pearson type III distribution. Chin (5) has given a table of frequency factors for probability levels suitable for analysis of minimum events such as droughts.

## Standard Error

*Method of Moments* - The general equation for the standard error of estimate of a 3-parameter distribution has been derived earlier as

$$S_T^2 = \frac{\mu_2}{n}\{1 + K\gamma_1 + \frac{K^2}{4}[\gamma_2 - 1] + \frac{\partial K}{\partial \gamma_1}[2\gamma_2 - 3\gamma_1^2 - 6$$

$$+ K(\gamma_3 - 6\gamma_1\gamma_2/4 - 10\gamma_1/4)] + \left(\frac{\partial K}{\partial \gamma_1}\right)^2 [\gamma_4$$

$$- 3\gamma_3\gamma_1 - 6\gamma_2 + 9\gamma_1^2\gamma_2/4 + 35\gamma_1^2/4 + 9]\} \qquad (9\text{-}53)$$

Substituting the relationships between the moment ratios of the Pearson type III distribution (Equations 9-19 - 9-21) this expression simplifies to

Table 9-2

Frequency Factor for Use in Pearson Type III Distribution

| Coefficient of Skew $\gamma_1$ | Cumulative Probability, P, % 50 | 80 | 90 | 95 | 98 | 99 |
|---|---|---|---|---|---|---|
| | Corresponding Return Period, T, Years 2 | 5 | 10 | 20 | 50 | 100 |
| 0.0 | 0.0000 | .8416 | 1.2816 | 1.6448 | 2.0537 | 2.3264 |
| 0.1 | -.0167 | .8363 | 1.2917 | 1.6728 | 2.1070 | 2.3997 |
| 0.2 | -.0333 | .8303 | 1.3009 | 1.6996 | 2.1595 | 2.4727 |
| 0.3 | -.0499 | .8234 | 1.3089 | 1.7254 | 2.2112 | 2.5453 |
| 0.4 | -.0664 | .8157 | 1.3159 | 1.7501 | 2.2619 | 2.6172 |
| 0.5 | -.0828 | .8072 | 1.3218 | 1.7735 | 2.3117 | 2.6884 |
| 0.6 | -.0990 | .7980 | 1.3267 | 1.7958 | 2.3603 | 2.7588 |
| 0.7 | -.1151 | .7880 | 1.3304 | 1.8168 | 2.4078 | 2.8283 |
| 0.8 | -.1310 | .7773 | 1.3330 | 1.8366 | 2.4541 | 2.8968 |
| 0.9 | -.1467 | .7659 | 1.3345 | 1.8551 | 2.4991 | 2.9641 |
| 1.0 | -.1621 | .7537 | 1.3349 | 1.8723 | 2.5428 | 3.0303 |
| 1.1 | -.1772 | .7409 | 1.3342 | 1.8881 | 2.5851 | 3.0952 |
| 1.2 | -.1921 | .7275 | 1.3324 | 1.9026 | 2.6260 | 3.1588 |
| 1.3 | -.2067 | .7134 | 1.3295 | 1.9157 | 2.6653 | 3.2209 |
| 1.4 | -.2209 | .6987 | 1.3255 | 1.9274 | 2.7031 | 3.2816 |
| 1.5 | -.2347 | .6834 | 1.3204 | 1.9378 | 2.7394 | 3.3406 |
| 1.6 | -.2482 | .6676 | 1.3143 | 1.9467 | 2.7740 | 3.3981 |
| 1.7 | -.2612 | .6513 | 1.3072 | 1.9543 | 2.8070 | 3.4538 |
| 1.8 | -.2738 | .6344 | 1.2990 | 1.9604 | 2.8383 | 3.5078 |
| 1.9 | -.2860 | .6171 | 1.2897 | 1.9651 | 2.8678 | 3.5600 |
| 2.0 | -.2977 | .5993 | 1.2795 | 1.9684 | 2.8956 | 3.6103 |

$$S_T^2 = \frac{\mu_2}{n}\{1 + K\gamma_1 + \frac{K^2}{2}[3\gamma_1^2/4 + 1] + 3K\frac{\partial K}{\partial \gamma_1}[\gamma_1 + \gamma_1^3/4]$$

$$+ 3(\frac{\partial K}{\partial \gamma_1})^2 [2 + 3\gamma_1^2 + 5\gamma_1^4/8]\} \qquad (9\text{-}54)$$

and now from Equation 9-52 the partial derivative of K with respect to $\gamma_1$ is

$$\frac{\partial K}{\partial \gamma_1} \approx \frac{t^2-1}{6} + \frac{4(t^3-6t)}{6^3}\gamma_1 - \frac{3(t^2-1)}{6^3}\gamma_1^2 + \frac{4t}{6^4}\gamma_1^3 - \frac{10}{6^6}\gamma_1^4 \qquad (9\text{-}55)$$

where t is the standard normal deviate. Substituting Equation 9-55 into Equation 9-54 and simplifying

$$S_T \approx \delta\sqrt{\frac{\mu_2}{n}} \qquad (9\text{-}56)$$

where $\delta$ is now dependent only on the required return period, T, and the sample coefficient of skew, $\gamma_1$. Table 9-3 gives values of $\delta$ for some values of T and $\gamma_1$. Bobée (2) has derived Equation 9-54 in a slightly different form.

Table 9-3

Parameter $\delta$ for Use in Standard Error of Pearson Type III Distribution

| Coefficient of skew $\gamma_1$ | Cumulative Probability, P, % | | | | | |
|---|---|---|---|---|---|---|
| | 50 | 80 | 90 | 95 | 98 | 99 |
| | Corresponding Return Period, T, Years | | | | | |
| | 2 | 5 | 10 | 20 | 50 | 100 |
| 0.0 | 1.0801 | 1.1698 | 1.3748 | 1.6845 | 2.1988 | 2.6363 |
| 0.1 | 1.0808 | 1.2006 | 1.4367 | 1.7810 | 2.3425 | 2.8168 |
| 0.2 | 1.0830 | 1.2309 | 1.4989 | 1.8815 | 2.4986 | 3.0175 |
| 0.3 | 1.0866 | 1.2609 | 1.5610 | 1.9852 | 2.6656 | 3.2365 |
| 0.4 | 1.0918 | 1.2905 | 1.6227 | 2.0915 | 2.8423 | 3.4724 |
| 0.5 | 1.0987 | 1.3199 | 1.6838 | 2.1998 | 3.0277 | 3.7238 |
| 0.6 | 1.1073 | 1.3492 | 1.7441 | 2.3094 | 3.2209 | 3.9895 |
| 0.7 | 1.1179 | 1.3785 | 1.8032 | 2.4198 | 3.4208 | 4.2684 |
| 0.8 | 1.1304 | 1.4082 | 1.8609 | 2.5303 | 3.6266 | 4.5595 |
| 0.9 | 1.1449 | 1.4385 | 1.9170 | 2.6403 | 3.8374 | 4.8618 |
| 1.0 | 1.1614 | 1.4699 | 1.9714 | 2.7492 | 4.0522 | 5.1741 |
| 1.1 | 1.1799 | 1.5030 | 2.0240 | 2.8564 | 4.2699 | 5.4952 |
| 1.2 | 1.2003 | 1.5382 | 2.0747 | 2.9613 | 4.4896 | 5.8240 |
| 1.3 | 1.2223 | 1.5764 | 2.1237 | 3.0631 | 4.7100 | 6.1592 |
| 1.4 | 1.2457 | 1.6181 | 2.1711 | 3.1615 | 4.9301 | 6.4992 |
| 1.5 | 1.2701 | 1.6643 | 2.2173 | 3.2557 | 5.1486 | 6.8427 |
| 1.6 | 1.2952 | 1.7157 | 2.2627 | 3.3455 | 5.3644 | 7.1881 |
| 1.7 | 1.3204 | 1.7732 | 2.3081 | 3.4303 | 5.5761 | 7.5339 |
| 1.8 | 1.3452 | 1.8374 | 2.3541 | 3.5100 | 5.7827 | 7.8783 |
| 1.9 | 1.3690 | 1.9091 | 2.4018 | 3.5844 | 5.9829 | 8.2196 |
| 2.0 | 1.3913 | 1.9888 | 2.4525 | 3.6536 | 6.1755 | 8.5562 |

*Maximum Likelihood* - As given earlier in Chapter 3 the variance of the T-year event, $S_T^2$, is expressed by maximum likelihood in terms of the parameter as

$$S_T^2 = \left(\frac{\partial x}{\partial \alpha}\right)^2 \text{var } \alpha + \left(\frac{\partial x}{\partial \beta}\right)^2 \text{var } \beta + \left(\frac{\partial x}{\partial \gamma}\right)^2 \text{var } \gamma + 2\frac{\partial x}{\partial \alpha}\frac{\partial x}{\partial \beta}\text{cov }(\alpha,\beta) + 2\frac{\partial x}{\partial \alpha}\frac{\partial x}{\partial \gamma}\text{cov }(\alpha,\gamma) + 2\frac{\partial x}{\partial \beta}\frac{\partial x}{\partial \gamma}\text{cov }(\beta,\gamma) \tag{9-57}$$

From the logarithm of the likelihood equation

$$\ln L = -n \ln \Gamma(\beta) - \frac{1}{\alpha}\sum_{i=1}^{n}(x_i-\gamma) + (\beta-1)\sum_{i=1}^{n}\ln(x_i-\gamma) - n\beta\ln\alpha \tag{9-58}$$

the following derivatives are obtained:

$$\frac{\partial^2 \ln L}{\partial \alpha^2} = \frac{-2 \sum_{i=1}^{n} (x_i - \gamma)}{\alpha^3} + \frac{n\beta}{\alpha^2} \tag{9-59}$$

$$\frac{\partial^2 \ln L}{\partial \beta^2} = - n\psi'(\beta) \tag{9-60}$$

where $\psi'(\beta)$ is the trigamma function.

The trigamma function, $\psi'(\beta)$, may be approximated by an asymptotic equation given by Abramowitz and Stegun (1). As with the psi function, this approximation is best combined with a recurrence equation to preserve accuracy at low $\beta$. A suitable equation is

$$\psi'(\beta) = \frac{1}{\beta+2} + \frac{1}{2(\beta+2)^2} + \frac{1}{6(\beta+2)^3} - \frac{1}{30(\beta+2)^5}$$

$$+ \frac{1}{42(\beta+2)^7} - \frac{1}{30(\beta+2)^9} + \frac{1}{(\beta+1)^2} + \frac{1}{\beta^2} \tag{9-61}$$

The remaining derivatives are:

$$\frac{\partial^2 \ln L}{\partial \gamma^2} = - (\beta-1) \sum_{i=1}^{n} (x_i-\gamma)^{-2} \tag{9-62}$$

$$\frac{\partial^2 \ln L}{\partial \alpha \partial \beta} = - \frac{n}{\alpha} \tag{9-63}$$

$$\frac{\partial^2 \ln L}{\partial \alpha \partial \gamma} = - \frac{n}{\alpha 2} \tag{9-64}$$

$$\frac{\partial^2 \ln L}{\partial \beta \partial \gamma} = - \sum_{i=1}^{n} (x_i-\gamma)^{-1} \tag{9-65}$$

The expression $\sum_{i=1}^{n} (x_i-\gamma)^r$ can be further evaluated from the pdf of the reduced variate, y,

$$f(x) = \frac{1}{\Gamma(\beta)} \int_o^\infty y^{\beta-1} e^{-y} dy \tag{9-66}$$

where $y = (x-\gamma)/\alpha$. If $(x_i-\gamma)^r$ is now introduced into the expression for moments about $\gamma$:

$$\mu_r^\gamma = \frac{1}{\Gamma(\beta)} \int_o^\infty (x_i - \gamma)^r y^{\beta-1} e^{-y} dy \tag{9-67}$$

$$= \frac{\alpha^r}{\Gamma(\beta)} \int_0^\infty y^{\beta+r-1} e^{-y} dy \qquad (9\text{-}68)$$

$$= \alpha^r \Gamma(\beta+r)/\Gamma(\beta) \qquad (9\text{-}69)$$

so that

$$\sum_{i=1}^{n} (x_i-\gamma)^r = n\mu_r^\gamma = n\alpha^r \Gamma(\beta+r)/\Gamma(\beta) \qquad (9\text{-}70)$$

Utilizing this expression for appropriate values of r the derivatives of the likelihood equation are reduced to

$$\frac{\partial^2 \ln L}{\partial \alpha^2} = - \frac{n\beta}{\alpha^2} \qquad (9\text{-}71)$$

$$\frac{\partial^2 \ln L}{\partial \beta^2} = - n\psi'(\beta) \qquad (9\text{-}72)$$

$$\frac{\partial^2 \ln L}{\partial \gamma^2} = - \frac{n}{\alpha^2(\beta-2)} \qquad (9\text{-}73)$$

$$\frac{\partial^2 \ln L}{\partial \alpha \partial \beta} = - \frac{n}{\alpha} \qquad (9\text{-}74)$$

$$\frac{\partial^2 \ln L}{\partial \alpha \partial \gamma} = = - \frac{n}{\alpha^2} \qquad (9\text{-}75)$$

$$\frac{\partial^2 \ln L}{\partial \beta \partial \gamma} = \frac{-n}{\alpha(\beta-1)} \qquad (9\text{-}76)$$

The likelihood or information matrix is then

$$I = n \begin{bmatrix} \frac{\beta}{\alpha^2} & \frac{1}{\alpha} & \frac{1}{\alpha^2} \\ & \psi'(\beta) & \frac{1}{\alpha(\beta-1)} \\ & & \frac{1}{\alpha^2(\beta-2)} \end{bmatrix} \qquad (9\text{-}77)$$

The variance and covariances of the parameters are obtained as the inverse of this matrix as:

$$\text{var } \alpha = \frac{1}{n\alpha^2 D}\left\{\frac{\psi'(\beta)}{(\beta-2)} - \frac{1}{(\beta-1)^2}\right\} \tag{9-78}$$

$$\text{var } \beta = \frac{2}{nD\alpha^4(\beta-2)} \tag{9-79}$$

$$\text{var } \gamma = \frac{\beta\psi'(\beta) - 1}{n\alpha^2 D} \tag{9-80}$$

$$\text{cov }(\alpha,\beta) = \frac{-1}{n\alpha^3 D}\left\{\frac{1}{(\beta-2)} - \frac{1}{(\beta-1)}\right\} \tag{9-81}$$

$$\text{cov }(\alpha,\gamma) = \frac{1}{n\alpha^2 D}\left[\frac{1}{(\beta-1)} - \psi'(\beta)\right] \tag{9-82}$$

$$\text{cov }(\beta,\gamma) = \frac{-1}{n\alpha^3 D}\left[\frac{\beta}{(\beta-1)} - 1\right] \tag{9-83}$$

where D is the determinant of the matrix of likelihood derivatives

$$D = \frac{1}{(\beta-2)\alpha^4}\left[2\psi'(\beta) - \frac{(2\beta-3)}{(\beta-1)^2}\right] \tag{9-84}$$

The partial derivatives needed can be obtained from 9-49 as:

$$\frac{\partial x}{\partial \alpha} = \left[\beta^{1/3} - \frac{1}{9\beta^{2/3}} + \frac{t}{3\beta^{1/6}}\right]^3 \tag{9-85}$$

$$\frac{\partial x}{\partial \beta} = 3\alpha\left[\beta^{1/3} - \frac{1}{9\beta^{2/3}} + \frac{t}{3\beta^{1/6}}\right]^2\left[\frac{1}{9\beta^{2/3}} + \frac{2}{27\beta^{5/3}} - \frac{t}{18\beta^{7/6}}\right] \tag{9-86}$$

$$\frac{\partial x}{\partial \gamma} = 1 \tag{9-87}$$

so that expression 9-57 can now be evaluated.

The maximum likelihood standard error of estimate for the two-parameter gamma distribution has been described by Moran (11) and Santos (12).

*Other Methods* - Bobée and Morin (3) have used order statistics to derive confidence intervals for the Pearson type III distribution. Tables and graphs show values of the standardized variable for given sample size, sample coefficient of skew, probability of exceedence (as an order statistic) and confidence level. The equivalent to this type of table using the standard error of estimate would be a modification of Equation 9-54 yielding confidence limits as

$$CL_{1,2} \approx \mu + K\sigma \pm \frac{t\ \delta\ \sigma}{\sqrt{n}} \qquad (9\text{-}88)$$

standardizing, this becomes

$$CL_{1,2} \approx K \pm \frac{t\ \delta}{\sqrt{n}} \qquad (9\text{-}89)$$

where K and $\delta$ are from Tables 9-2 and 9-3, t is the standard normal deviate for the required confidence level and n is the sample size. Values from Equation 9-89 approximate those given by Bobée and Morin.

## References


1. Abramowitz, M. and I. A. Stegun, 1965, Handbook of Mathematical Functions, Dover Publications, New York.

2. Bobée, B, 1973, Sample Error of T-year Events Computed by Fitting a Pearson Type 3 Distribution, Wat. Res. Res., Vol. 9, No. 5, pp. 1264-1270.

3. Bobée, B. and G. Morin, 1973, Statistique D'ordre de la Loi Pearson III et de sa Forme Derivée à Asymétrie Négative, Revue de Statistique Appliquee, Vol. XXI, No. 4, pp. 69-80.

4. Bobée, B. and R. Robitaille, 1976, The Use of the Pearson Type 3 and Log-Pearson Type 3 Distributions Revisited, Paper submitted to Wat. Res. Res.

5. Chin, W. O., 1967, Formulae and Tables for Computing and Plotting Drought Frequency Curves, Technical Bulletin No. 8, Inland Waters Branch, Ottawa.

6. Condie, R. and G. Nix, 1975, Modelling of Low Flow Frequency Distributions and Parameter Estimation, Int. Wat. Res. Assoc., Proc. Symp. on Water for Arid Lands, Tehran, Dec. 8 and 9.

7. Greenwood, J. A. and D. Durand, 1960, Aids for Fitting the Gamma Distribution, Technometrics, Vol. 2, No. 1, pp. 55-66.

8. Kendall, M. G. and A. Stuart, 1963, The Advanced Theory of Statistics, Vol. I, Griffin, London.

9. Matalas, N. C., 1963, Probability Distribution of Low Flows, USGS Professional Paper No. 434-A.

10. Matalas, N. C. and J. R. Wallis, 1973, Eureka! It Fits a Pearson Type 3 Distribution, Wat. Res. Res., Vol. 9, No. 2, pp. 281-289.

11. Moran, P. A. P., 1957, The Statistical Treatment of Flood Flows, Trans. American Geophysical Union, Vol. 38, No. 4, pp. 519-523.

12. Santos, A., 1970, The Statistical Treatment of Flood Flows, Water Power, Vol. 22, No. 2, pp. 63-67.

13. Soil Conservation Service, U. S. Dept. of Agriculture, 1968, New Tables of Percentage Points of the Pearson Type III Distribution, Technical Release No. 38, Central Technical Unit.

```
      PROGRAM PT3(INPUT,OUTPUT,TAPE5=INPUT,TAPE6=OUTPUT)
C
C
C     COMPUTES METHOD OF MOMENTS AND MAXIMUM LIKELIHOOD ESTIMATES FOR
C     T YEAR EVENTS AND STANDARD ERRORS FOR PEARSON TYPE 3
C     DISTRIBUTION
C     INPUT
C     TITLE
C     N NUMBER OF ANNUAL MAXIMUM EVENTS
C     X SERIES OF EVENTS
C
C
      REAL M1,M2,M3,K
      DIMENSION SND(6), X(100)
      DIMENSION XT(6), SX(6), TITLE(80)
      SND(1)=0.
      SND(2)=0.84162
      SND(3)=1.28155
      SND(4)=1.64485
      SND(5)=2.05375
      SND(6)=2.32635
      READ (5,9) TITLE
      READ (5,10) N
      READ (5,11) (X(I),I=1,N)
      XN=N
      A=0.0
      B=0.0
      C=0.0
      DO 1 I=1,N
      A=A+X(I)
      B=B+X(I)**2
      C=C+X(I)**3
1     CONTINUE
      M1=A/XN
      M2=(B/XN)-(A/XN)**2
      M3=(C/XN)+2.0*M1**3-3.0*M1*(B/XN)
      SKEW=M3/(M2**1.5)
      C1=(SQRT(XN*(XN-1.0)))/(XN-2.0)
      C2=1.0+8.5/XN
      C3=XN/(XN-1.0)
      SKEW=SKEW*C1*C2
      M2=M2*C3
      BETA=(2.0/SKEW)**2
      ALPHA=(M2**0.5)/(BETA**0.5)
      GAMMA=M1-(M2**0.5)*(BETA**0.5)
      WRITE (6,12) TITLE
      WRITE (6,13)
      WRITE (6,22) ALPHA,M1
      WRITE (6,23) BETA,M2
      WRITE (6,24) GAMMA,SKEW
      DO 2 J=1,6
      T=SND(J)
      T1=T
      T2=(T**2-1.0)/6.0
      T3=2.0*(T**3-6.0*T)/6.0**3
      T4=(T**2-1.0)/6.0**3
      T5=T/6.0**4
      T6=2.0/6.0**6
      K=T1+T2*SKEW+T3*SKEW**2-T4*SKEW**3+T5*SKEW**4-T6*SKEW**5
      SLOPE=T2+T3*2.0*SKEW-T4*3.0*SKEW**2+T5*4.0*SKEW**3-T6*5.0*SKEW**4
      T7=(1.0+0.75*SKEW**2)*(0.5*K**2)
      T8=K*SKEW
      T9=6.0*(1.0+0.25*SKEW**2)*SLOPE
      T10=SLOPE*(1.0+1.25*SKEW**2)+(SKEW*K/2.0)
      DELTA=T7+T8+T9*T10
      XT(J)=M1+K*SQRT(M2)
2     SX(J)=SQRT(M2*DELTA/XN)
      WRITE (6,14)
      WRITE (6,15) (XT(J),J=1,6)
      WRITE (6,16) (SX(J),J=1,6)
      WRITE (6,17)
      WRITE (6,18)
      ICOUNT=0
      XMIN=10000000.
      DO 3 I=1,N
3     IF (X(I).LT.XMIN) XMIN=X(I)
      GML=XMIN*0.99
4     ICOUNT=ICOUNT+1
      A=0.0
      B=0.0
```

```
      C=0.0
      R=0.0
      DO 5 I=1,N
      A=A+1.0/(X(I)-GML)
      B=B+(X(I)-GML)
      C=C+ALOG(X(I)-GML)
      R=R+1.0/((X(I)-GML)**2)
5     CONTINUE
      BETA=A/(A-(XN**2)/B)
      ALPHA=B/(XN*BETA)
      D=BETA+2.0
      PSI=ALOG(D)-(1.0/(2.0*D))-(1.0/(12.0*D**2))+(1.0/(120.0*D**4))-(1.
     10/(252.0*D**6))-(1.0/(BETA+1.0))-(1.0/BETA)
      FCN=-XN*PSI+C-XN*ALOG(ALPHA)
      TRI=(1.0/D)+(1.0/(2.0*D**2))+(1.0/(6.0*D**3))-(1.0/(30.0*D**5))+(1
     1.0/(42.0*D**7))-(1.0/(30.0*D**9))+(1.0/((BETA+1.0)**2))+(1.0/(BETA
     2**2))
      V=A-(XN**2)/B
      U=A
      W=(B/XN)-(XN/A)
      DU=R
      DV=R-(XN**3)/(B**2)
      DW=-1.0+(XN*R)/(A**2)
      FPN=-XN*TRI*((V*DU-U*DV)/(V**2))-A-XN*DW/W
      AS=GML-(FCN/FPN)
      WRITE (6,19) ICOUNT,AS,FCN
      DELTA=ABS(0.00000001*AS)
      IF (ABS(AS-GML).LT.DELTA) GO TO 6
      IF (ICOUNT.GT.25) GO TO 8
      GML=AS
      GO TO 4
6     CONTINUE
      GAMMA=AS
      M1=GAMMA+ALPHA*BETA
      M2=BETA*ALPHA**2
      SKEW=2.0/SQRT(BETA)
      WRITE (6,20)
      WRITE (6,22) ALPHA,M1
      WRITE (6,23) BETA,M2
      WRITE (6,24) GAMMA,SKEW
      WRITE (6,20)
      D=BETA+2.0
      TRI=(1.0/D)+(1.0/(2.0*D**2))+(1.0/(6.0*D**3))-(1.0/(30.0*D**5))+(1
     1.0/(42.0*D**7))-(1.0/(30.0*D**9))+(1.0/((BETA+1.0)**2))+(1.0/(BETA
     2**2))
      H=(BETA-2.0)*ALPHA**4
      P=2.0*TRI-(2.0*BETA-3.0)/((BETA-1.0)**2)
      DET=P/H
      VARA=(1.0/(XN*(ALPHA**2)*DET))*((TRI/(BETA-2.0))-1.0/((BETA-1.0)**
     12))
      VARB=2.0/(XN*DET*(BETA-2.0)*ALPHA**4)
      VARG=(BETA*TRI-1.0)/(XN*DET*ALPHA**2)
      COVAB=(-1.0/(XN*DET*ALPHA**3))*((1.0/(BETA-2.0))-(1.0/(BETA-1.0)))
      COVAG=(1.0/(XN*DET*ALPHA**2))*((1.0/(BETA-1.0))-TRI)
      COVBG=(-1.0/(XN*DET*ALPHA**3))*((BETA/(BETA-1.0))-1.0)
      DO 7 J=1,6
      T=SND(J)
      E=BETA**(1./3.)-1.0/(9.0*BETA**(2./3.))+T/(3.0*BETA**(1./6.))
      F=1.0/(3.0*BETA**(2./3.))+2.0/(27.0*BETA**(5./3.))-T/(18.0*BETA**(
     17./6.))
      XT(J)=GAMMA+ALPHA*E**3
      DXDA=E**3
      DXDB=3.0*ALPHA*E**2*F
      DXDG=1.0
      SX(J)=SQRT(VARA*DXDA**2+VARB*DXDB**2+VARG*DXDG**2+2.0*DXDA*DXDB*CO
     1VAB+2.0*DXDA*DXDG*COVAG+2.0*DXDB*DXDG*COVBG)
7     CONTINUE
      WRITE (6,14)
      WRITE (6,15) (XT(J),J=1,6)
      WRITE (6,16) (SX(J),J=1,6)
      WRITE (6,21)
8     CONTINUE
      STOP
C
C
9     FORMAT (80A1)
10    FORMAT (I5)
11    FORMAT (8F10.0)
12    FORMAT (1H1,/,80A1,//,26X,28HPEARSON TYPE 3 DISTRIBUTION ,/)
13    FORMAT (31X,17HMETHOD OF MOMENTS,//)
```

```
14    FORMAT (3X,7HT,YEARS,4X,1H2,11X,1H5,10X,2H10,10X,2H20,10X,2H50,9X,
     13H100,/)
15    FORMAT (3X,1HX,3X,6E12.5,/,4X,1HT)
16    FORMAT (3X,1HS,3X,6E12.5,/,4X,1HT,//)
17    FORMAT (25X,28HMAXIMUM LIKELIHOOD PROCEDURE,//)
18    FORMAT (21X,5HTRIAL,11X,1HG,11X,4HF(G),/)
19    FORMAT (22X,I2,8X,E12.5,1X,E12.5)
20    FORMAT (//)
21    FORMAT (/1H1)
22    FORMAT (9X,5HALPHA,5X,E12.5,14X,4HM1   ,6X,E12.5)
23    FORMAT (9X,5HBETA ,5X,E12.5,14X,4HM2   ,6X,E12.5)
24    FORMAT (9X,5HGAMMA,5X,E12.5,14X,4HSKEW,6X,E12.5,/)
      END
```

ST. MARYS RIVER AT STILLWATER - STATION NO. 01E0001

PEARSON TYPE 3 DISTRIBUTION

METHOD OF MOMENTS

| | | | |
|---|---|---|---|
| ALPHA | .40451E+04 | M1 | .14555E+05 |
| BETA | .16697E+01 | M2 | .27320E+08 |
| GAMMA | .78007E+04 | SKEW | .15478E+01 |

| T,YEARS | 2 | 5 | 10 | 20 | 50 | 100 |
|---|---|---|---|---|---|---|
| $X_T$ | .13294E+05 | .18088E+05 | .21442E+05 | .24707E+05 | .28961E+05 | .32160E+05 |
| $S_T$ | .54136E+03 | .91782E+03 | .13518E+04 | .21215E+04 | .34793E+04 | .46803E+04 |

MAXIMUM LIKELIHOOD PROCEDURE

| TRIAL | G | F(G) |
|---|---|---|
| 1 | .65106E+04 | .74097E+01 |
| 2 | .63101E+04 | .34530E+01 |
| 3 | .60365E+04 | .15006E+01 |
| 4 | .57606E+04 | .57166E+00 |
| 5 | .55913E+04 | .17020E+00 |
| 6 | .55489E+04 | .29106E-01 |
| 7 | .55469E+04 | .12866E-02 |
| 8 | .55469E+04 | .27976E-05 |
| 9 | .55469E+04 | .16371E-10 |

| | | | |
|---|---|---|---|
| ALPHA | .29466E+04 | M1 | .14555E+05 |
| BETA | .30570E+01 | M2 | .26543E+08 |
| GAMMA | .55469E+04 | SKEW | .11439E+01 |

| T,YEARS | 2 | 5 | 10 | 20 | 50 | 100 |
|---|---|---|---|---|---|---|
| $X_T$ | .13608E+05 | .18342E+05 | .21425E+05 | .24316E+05 | .27966E+05 | .30646E+05 |
| $S_T$ | .63268E+03 | .98015E+03 | .13285E+04 | .17198E+04 | .22795E+04 | .27264E+04 |

# CHAPTER 10

## LOG-PEARSON TYPE III DISTRIBUTION

### Introduction

The U. S. Federal Water Resources Council recommended in 1967 that the log-Pearson type III distribution be adopted as the standard flood frequency distribution by all U. S. government agencies. In describing the investigations behind this recommendation, Benson (3) explained that no rigorous statistical criteria exist on which a comparison of distributions can be based and therefore the choice of the log-Pearson type III was, to some extent, subjective.

If the logarithms, ln x, of a variable x are distributed as a Pearson type III variate then the variable x will be distributed as a log-Pearson type III with probability density function

$$p(x) = \frac{1}{\alpha x \Gamma(\beta)} \left\{\frac{\ln x - \gamma}{\alpha}\right\}^{\beta-1} e^{-\left\{\frac{\ln x - \gamma}{\alpha}\right\}} \qquad (10\text{-}1)$$

where $\alpha$, $\beta$ and $\gamma$ are the scale, shape and location parameters respectively.

Being a 3-parameter distribution operating upon the logarithms of the variable the log-Pearson type III would appear to be an extremely versatile distribution. However, its applicability in hydrology is strictly limited. As shown by Bobée (1) the pdf of the log-Pearson type III may be J-shaped, reverse J-shaped, U-shaped, inverted U-shaped, inverted U with inflexions, bell shaped with an upper bound, bell shaped with a lower bound, etc, etc.. For flood frequency analysis the only shape of interest is that which is unimodal, continuous from 0 to $+\infty$, has either an infinitely high order ($\partial p/\partial x = \infty$) or smooth contact ($\partial p/\partial x = 0$) at the lower limit and is unbounded at the upper limit. The log-Pearson type III falls within these criteria only when $\beta > 1$ and when $1/\alpha > 0$. If the coefficient of skew, $\gamma$, is negative this corresponds to a negative value of $\alpha$ and is not suitable. Reich (5) has given examples of the application of the log-Pearson type III distribution to samples with negative skew in which the computed upper bounds have been lower than the maximum observed events.

### Estimation of Parameters

As with the lognormal distribution, two procedures are available to estimate the parameters of a log-Pearson type III distribution. Either the distribution can be fitted directly to the data or else the Pearson type III distribution can be fitted to the logarithms of the events. In the first case the method of moments or maximum lieklihood can be applied. In the second case the maximum likelihood method is the same as in the first case and only the method of moments gives new results. As with the lognormal distributions care must be taken with events of zero magnitude. Jennings and Benson (4) describe a procedure of coping with these events.

*Method of Moments (Direct Application of Log-Pearson type III)* - Applying the moment generating function to the pdf of the log-Pearson type III distribution yields, following Bobée (1)

$$\mu_r' = \frac{e^{\gamma r}}{(1 - r\alpha)^{\beta}} \tag{10-2}$$

Evaluating this expression for the first, second and third moments about the origin gives

$$\ln \mu_1' = \gamma - \beta \ln (1 - \alpha) \tag{10-3}$$

$$\ln \mu_2' = 2\gamma - \beta \ln (1 - 2\alpha) \tag{10-4}$$

$$\ln \mu_3' = 3\gamma - \beta \ln (1 - 3\alpha) \tag{10-5}$$

Manipulating these three equations gives

$$\frac{\ln \mu_3' - 3 \ln \mu_1'}{\ln \mu_2' - 2 \ln \mu_1'} = \frac{\ln [(1 - \alpha)^3/(1 - 3\alpha)]}{\ln [(1 - \alpha)^2/(1 - 2\alpha)]} \tag{10-6}$$

The left hand side of Equation 10-6, call it B,

$$B = \frac{\ln \mu_3' - 3 \ln \mu_1'}{\ln \mu_2' - 2 \ln \mu_1'} \tag{10-7}$$

can be calculated directly from the sample moments about the origin. It is then necessary to compute $\alpha$ from B. This is difficult analytically in Equation 10-6 but by substituting

$$A = 1/\alpha - 3 \tag{10-8}$$

and

$$C = 1/(B - 3) \tag{10-9}$$

a series of polynomials can be derived.

Figure 10-1 shows the shape of the function relating $\alpha$ to B. From this figure and noting the restrictions on the useful range of $\alpha$, it is only necessary to relate A and C for $3 < B \leq 6$. Two regressions have been developed to cover this range:

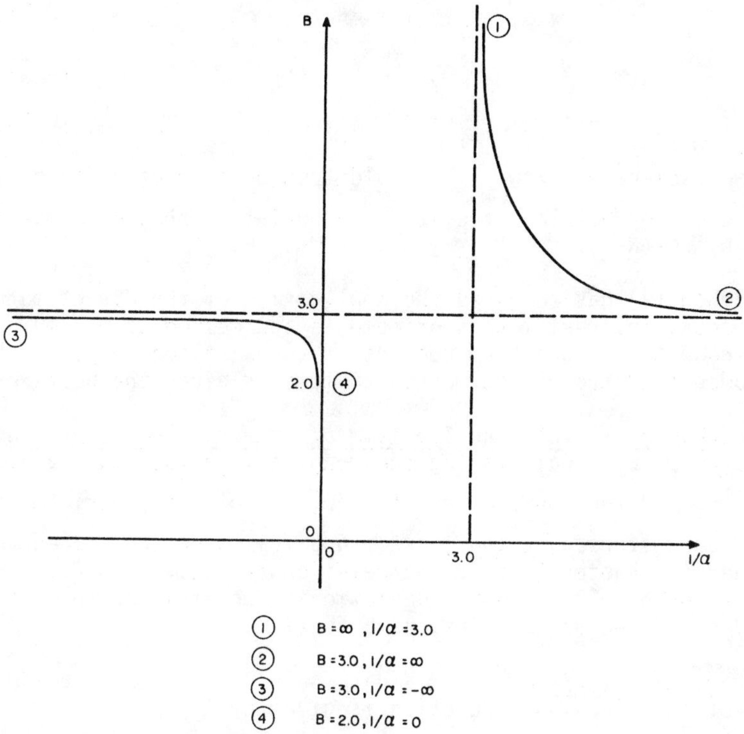


Figure 10-1. Relationship between α and B in equation 10-8. (not to scale)

(a) for $3.5 < B \leq 6.0$

$$A = -0.23019 + 1.65262C + 0.20911C^2 - 0.04557C^3 \qquad (10\text{-}10)$$

(b) for $3.0 < B \leq 3.5$

$$A = -\ 0.47157 + 1.99955C \qquad (10\text{-}11)$$

Having computed B and α the remaining parameters, β and γ, can be obtained from reorganizations of Equations 10-3 and 10-4 as

$$\beta = \frac{\ln \mu_2' - 2 \ln \mu_1'}{\ln [\ln (1 - \alpha)^2 - \ln (1 - 2\alpha)]} \qquad (10\text{-}12)$$

$$\gamma = \ln \mu_1' + \beta \ln (1 - \alpha) \qquad (10\text{-}13)$$

The parameters α, β and γ are then used to compute the mean, $\mu_y$, standard deviation, $\sigma_y$, and coefficient of skew, $\gamma_y$, of the logarithms of x as:

$$\mu_y = \gamma + \alpha\beta \tag{10-14}$$

$$\sigma_y = \alpha\sqrt{\beta} \tag{10-15}$$

and

$$\gamma_y = 2.0/\sqrt{\beta} \tag{10-16}$$

The coefficient of skew, $\gamma_y$, should then be corrected for bias as in Equation 9.22 and the standard deviation should be multiplied by $\sqrt{N/(N-1)}$.

Bobée (2) has compared the application of the direct method of moments, indirect method of moments and method of maximum likelihood to 18 long term records of annual flood peaks and has concluded that the direct method of moments gives the best results.

*Method of Moments (Application of Pearson type III to the Logarithm of x)* - Calculating the mean, $\mu_y$, the standard deviation, $\sigma_y$, and the coefficient of skew, $\gamma_y$, of the logarithms of x, the parameters $\alpha$, $\beta$ and $\gamma$ are found following the procedure outlined in Chapter 9 using the relationships given in Equations 10-14 - 10-16. The standard deviation and coefficient of skew should be corrected for bias, as before.

*Maximum Likelihood* - For a data sample $x_1,\ldots,x_n$ the logarithm of the likelihood function is given by

$$\ln L = -\sum_{i=1}^{n} \ln x - n \ln \Gamma(\beta) - \frac{\sum_{i=1}^{n} (\ln x_i - \gamma)}{\alpha}$$

$$+ (\beta - 1) \sum_{i=1}^{n} \ln (\ln x_i - \gamma) - n\beta \ln \alpha \tag{10-17}$$

Differentiating with respect to $\alpha$, $\beta$ and $\gamma$ and equating to zero yields:

$$\sum_{i=1}^{n} (\ln x_i - \alpha) = n\alpha\beta \tag{10-18}$$

$$n\psi(\beta) = \sum_{i=1}^{n} \ln [(\ln x - \gamma)/\alpha] \tag{10-19}$$

$$n = \alpha(\beta - 1) \sum_{i=1}^{n} 1/(\ln x_i - \gamma) \tag{10-20}$$

where, as before, $\psi(\beta)$ is the psi or digamma function which can be calculated by an asymptotic expansion.

Eliminating $\alpha$ in 10-20 by means by 10-18 yields an expression in x, $\beta$ and $\gamma$. Substitution in 10-19 leaves an expression in $\gamma$ only which can be solved numerically.

Frequency Factor

Applying the Pearson type III distribution to the logarithms of the sample events the T-year event can be computed from

$$y_T = \ln x_T = \mu_y + K\sigma_y \tag{10-21}$$

where $\mu_y$ and $\sigma_y$ are the mean and standard deviation of the logarithms of x and K is obtained from Table 9-2 or Equation 9-52 using $\gamma_y$, the coefficient of skew of the logarithms, in place of $\gamma_1$, the coefficient of skew of the events.

There is no need to derive a frequency factor without logarithmic components, as was done with the lognormal distribution, since the logarithmic moments can be derived directly from the moments of the observations using equations 10-14 - 10-16.

## Standard Error

*Method of Moments* - The standard error of the T-year event for the log-Pearson type III distribution may be computed using first Equation 9-56 to obtain $S_{T,y}$ in log units from the standard normal deviate and the coefficient of skew of the logarithms of the observed events.

The standard error is then converted back to linear units as

$$S_{T,x} = \frac{x_T \left(e^{S_{T,y}} - e^{-S_{T,y}}\right)}{2.0} \tag{10-22}$$

where $S_{T,x}$ is now the average of the positive and negative standard errors in linear units and $x_T$ is the T-year event.

*Maximum Likelihood* - The standard error in log units is obtained from Equation 9-57 and subsequent development using parameters $\alpha$, $\beta$ and $\gamma$ derived from section on Maximum Likelihood (p. 126). The logarithmic standard error is then converted to linear units using equation 10-22 above.

## Computer Program

There are two differences between the printed output from the log-Pearson type III program and those from the other programs:

(a) L1, L2 and L3 are the 1st, 2nd and 3rd moments of the original data about the origin, required in the direct method of mements.

(b) The values of M1, M2 and SKEW printed out for the indirect method of moments and method of maximum likelihood are moments of the logarithms of the sample.

References


1. Bobée, B., 1975, The Lob-Pearson Type III Distribution and its Application in Hydrology, Wat. Res. Res., Vol. 11, No. 5, pp. 681-689.

2. Bobée, B., 1976, The Use of the Pearson Type 3 and Log-Pearson Type 3 Distributions Revisited, Paper submitted to Wat. Res. Res.

3. Benson, M. A., 1968, Uniform Flood Frequency Estimating Methods for Federal Agencies, Wat. Res. Res., Vol. 4, No. 5, pp. 891-908.

4. Jennings, M. E. and M. A. Benson, 1969, Frequency Curves for Annual Flood Series With Some Zero Events or Incomplete Data, Wat. Res. Res., Vol. 5, No. 1, pp. 276-280.

5. Reich, B. M., 1972, Log-Pearson Type 3 and Gumbel Analysis of Floods, Second International Symposium in Hydrology, Fort Collins, Colorado, pp. 290-303.

```
      PROGRAM LP3(INPUT,OUTPUT,TAPE5=INPUT,TAPE6=OUTPUT)
C
C
C     COMPUTES METHOD OF MOMENTS AND MAXIMUM LIKELIHOOD ESTIMATES FOR
C     T YEAR EVENTS AND STANDARD ERRORS FOR LOG-PEARSON TYPE 3
C     DISTRIBUTION
C     INPUT
C     TITLE
C     N NUMBER OF ANNUAL MAXIMUM EVENTS
C     X SERIES OF EVENTS
C
C
      REAL M1,M2,M3,K,NSX,L1,L2,L3
      DIMENSION SND(6), X(100)
      DIMENSION XT(6), TITLE(80), ST(6)
      SND(1)=0.
      SND(2)=0.84162
      SND(3)=1.28155
      SND(4)=1.64485
      SND(5)=2.05375
      SND(6)=2.32635
      READ (5,16) TITLE
      READ (5,17) N
      XN=N
      READ (5,18) (X(I),I=1,N)
      C1=(SQRT(XN*(XN-1.0)))/(XN-2.0)
      C2=1.0+8.5/XN
      C3=XN/(XN-1.0)
      WRITE (6,19) TITLE
      WRITE (6,20)
      A=0.0
      B=0.0
      C=0.0
      DO 1 I=1,N
      A=A+X(I)
      B=B+X(I)**2
      C=C+X(I)**3
1     CONTINUE
      L1=A/XN
      L2=B/XN
      L3=C/XN
      M1=A/XN
      M2=(B/XN)-(A/XN)**2
      M3=(C/XN)+2.0*M1**3-3.0*M1*(B/XN)
      SKEW=M3/(M2**1.5)
      B=(ALOG(L3)-3.0*ALOG(L1))/(ALOG(L2)-2.0*ALOG(L1))
      WRITE (6,34) L1,M1
      WRITE (6,35) L2,M2
      WRITE (6,36) L3,SKEW
      C=1.0/(B-3.0)
      IF (B.GT.6.0) GO TO 3
      IF (B.LE.3.0) GO TO 3
      IF (B.LE.3.5) GO TO 2
      A=-0.23019+1.65262*C+0.20911*C**2-0.04557*C**3
      GO TO 4
2     A=-0.47157+1.99955*C
      GO TO 4
3     WRITE (6,21) B
      GO TO 6
4     ALPHA=1.0/(A+3.0)
      A1=ALOG(1.0-ALPHA)
      A2=ALOG(1.0-2.0*ALPHA)
      BETA=(ALOG(L2)-2.0*ALOG(L1))/(2.0*A1-A2)
      GAMMA=ALOG(L1)+BETA*A1
      M1=GAMMA+ALPHA*BETA
      M2=BETA*ALPHA**2
      SKEW=2.0/SQRT(BETA)
      WRITE (6,23) ALPHA,M1
      WRITE (6,24) BETA,M2
      WRITE (6,25) GAMMA,SKEW
      IF (SKEW.LT.0.0) WRITE (6,33)
      DO 5 J=1,6
      T=SND(J)
      T1=T
      T2=(T**2-1.0)/6.0
      T3=2.0*(T**3-6.0*T)/6.0**3
      T4=(T**2-1.0)/6.0**3
      T5=T/6.0**4
      T6=2.0/6.0**6
      K=T1+T2*SKEW+T3*SKEW**2-T4*SKEW**3+T5*SKEW**4-T6*SKEW**5
```

```
      SLOPE=T2+T3*2.0*SKEW-T4*3.0*SKEW**2+T5*4.0*SKEW**3-T6*5.0*SKEW**4
      T7=1.0
      T8=SKEW*K
      T9=(1.0+0.75*SKEW**2)*((K**2)/2.0)
      T10=3.0*SLOPE*K*(SKEW+0.25*SKEW**3)
      T11=3.0*(SLOPE**2)*(2.0+3.0*SKEW**2+(5.0/8.0)*SKEW**4)
      DELTA=T7+T8+T9+T10+T11
      XT(J)=EXP(M1+K*SQRT(M2))
      SX=SQRT(M2*DELTA/XN)
      PSX=XT(J)*(EXP(SX)-1.0)
      NSX=-XT(J)*(EXP(-SX)-1.0)
      ST(J)=(PSX+NSX)/2.0
5     CONTINUE
      WRITE (6,26)
      WRITE (6,27) (XT(J),J=1,6)
      WRITE (6,28) (ST(J),J=1,6)
6     DO 7 I=1,N
7     X(I)=ALOG(X(I))
      A=0.0
      B=0.0
      C=0.0
      DO 8 I=1,N
      A=A+X(I)
      B=B+X(I)**2
      C=C+X(I)**3
8     CONTINUE
      M1=A/XN
      M2=(B/XN)-(A/XN)**2
      M3=(C/XN)+2.0*M1**3-3.0*M1*(B/XN)
      SKEW=M3/(M2**1.5)
      SKEW=SKEW*C1*C2
      M2=M2*C3
      BETA=(2.0/SKEW)**2
      ALPHA=(M2**0.5)/(BETA**0.5)
      GAMMA=M1-(M2**0.5)*(BETA**0.5)
      WRITE (6,22)
      WRITE (6,23) ALPHA,M1
      WRITE (6,24) BETA,M2
      WRITE (6,25) GAMMA,SKEW
      IF (SKEW.LT.0.0) WRITE (6,33)
      DO 9 J=1,6
      T=SND(J)
      T1=T
      T2=(T**2-1.0)/6.0
      T3=2.0*(T**3-6.0*T)/6.0**3
      T4=(T**2-1.0)/6.0**3
      T5=T/6.0**4
      T6=2.0/6.0**6
      K=T1+T2*SKEW+T3*SKEW**2-T4*SKEW**3+T5*SKEW**4-T6*SKEW**5
      SLOPE=T2+T3*2.0*SKEW-T4*3.0*SKEW**2+T5*4.0*SKEW**3-T6*5.0*SKEW**4
      T7=1.0
      T8=SKEW*K
      T9=(1.0+0.75*SKEW**2)*((K**2)/2.0)
      T10=3.0*SLOPE*K*(SKEW+0.25*SKEW**3)
      T11=3.0*(SLOPE**2)*(2.0+3.0*SKEW**2+(5.0/8.0)*SKEW**4)
      DELTA=T7+T8+T9+T10+T11
      XT(J)=EXP(M1+K*SQRT(M2))
      SX=SQRT(M2*DELTA/XN)
      PSX=XT(J)*(EXP(SX)-1.0)
      NSX=-XT(J)*(EXP(-SX)-1.0)
      ST(J)=(PSX+NSX)/2.0
9     CONTINUE
      WRITE (6,26)
      WRITE (6,27) (XT(J),J=1,6)
      WRITE (6,28) (ST(J),J=1,6)
      WRITE (6,29)
      WRITE (6,30)
      ICOUNT=0
      XMIN=10000000.
      DO 10 I=1,N
10    IF (X(I).LT.XMIN) XMIN=X(I)
      GML=XMIN*0.99
11    ICOUNT=ICOUNT+1
      A=0.0
      B=0.0
      C=0.0
      R=0.0
      DO 12 I=1,N
      A=A+1.0/(X(I)-GML)
      B=B+(X(I)-GML)
```

```
      C=C+ALOG(X(I)-GML)
      R=R+1.0/((X(I)-GML)**2)
12    CONTINUE
      BETA=1.0/(1.0-(XN**2)/(B*A))
      ALPHA=(B/XN)-(XN/A)
      D=BETA+2.0
      PSI=ALOG(D)-(1.0/(2.0*D))-(1.0/(12.0*D**2))+(1.0/(120.0*D**4))-(1.
     10/(252.0*D**6))-(1.0/(BETA+1.0))-(1.0/BETA)
      FCN=-XN*PSI+C-XN*ALOG(ALPHA)
      TRI=(1.0/D)+(1.0/(2.0*D**2))+(1.0/(6.0*D**3))-(1.0/(30.0*D**5))+(1
     1.0/(42.0*D**7))-(1.0/(30.0*D**9))+(1.0/((BETA+1.0)**2))+(1.0/(BETA
     2**2))
      V=A-(XN**2)/B
      U=A
      W=(B/XN)-(XN/A)
      DU=R
      DV=R-(XN**3)/(B**2)
      DW=-1.0+(XN*R)/(A**2)
      FPN=-XN*TRI*((V*DU-U*DV)/(V**2))-A-XN*DW/W
      AS=GML-(FCN/FPN)
      WRITE (6,31) ICOUNT,AS,FCN
      DELTA=ABS(0.00000001*AS)
      IF (ABS(AS-GML).LT.DELTA) GO TO 13
      IF (ICOUNT.GE.25) GO TO 15
      GML=AS
      GO TO 11
13    CONTINUE
      GAMMA=AS
      M1=GAMMA+ALPHA*BETA
      M2=BETA*ALPHA**2
      SKEW=2.0/SQRT(BETA)
      WRITE (6,32)
      WRITE (6,23) ALPHA,M1
      WRITE (6,24) BETA,M2
      WRITE (6,25) GAMMA,SKEW
      IF (SKEW.LT.0.0) WRITE (6,33)
      D=BETA+2.0
      TRI=(1.0/D)+(1.0/(2.0*D**2))+(1.0/(6.0*D**3))-(1.0/(30.0*D**5))+(1
     1.0/(42.0*D**7))-(1.0/(30.0*D**9))+(1.0/((BETA+1.0)**2))+(1.0/(BETA
     2**2))
      H=(BETA-2.0)*ALPHA**4
      P=2.0*TRI-(2.0*BETA-3.0)/((BETA-1.0)**2)
      DET=P/H
      VARA=(1.0/(XN*(ALPHA**2)*DET))*((TRI/(BETA-2.0))-1.0/((BETA-1.0)**
     12))
      VARB=2.0/(XN*DET*(BETA-2.0)*ALPHA**4)
      VARG=(BETA*TRI-1.0)/(XN*DET*ALPHA**2)
      COVAB=(-1.0/(XN*DET*ALPHA**3))*((1.0/(BETA-2.0))-(1.0/(BETA-1.0)))
      COVAG=(1.0/(XN*DET*ALPHA**2))*((1.0/(BETA-1.0))-TRI)
      COVBG=(-1.0/(XN*DET*ALPHA**3))*((BETA/(BETA-1.0))-1.0)
      DO 14 J=1,6
      T=SND(J)
      E=BETA**(1./3.)-1.0/(9.0*BETA**(2./3.))+T/(3.0*BETA**(1./6.))
      F=1.0/(3.0*BETA**(2./3.))+2.0/(27.0*BETA**(5./3.))-T/(18.0*BETA**(
     17./6.))
      XT(J)=EXP(ALPHA*E**3+GAMMA)
      DXDA=E**3
      DXDB=3.0*ALPHA*E**2*F
      DXDG=1.0
      SX=SQRT(VARA*DXDA**2+VARB*DXDB**2+VARG*DXDG**2+2.0*DXDA*DXDB*COVAB
     1+2.0*DXDA*DXDG*COVAG+2.0*DXDB*DXDG*COVBG)
      PSX=XT(J)*(EXP(SX)-1.0)
      NSX=-XT(J)*(EXP(-SX)-1.0)
      ST(J)=(PSX+NSX)/2.0
14    CONTINUE
      WRITE (6,26)
      WRITE (6,27) (XT(J),J=1,6)
      WRITE (6,28) (ST(J),J=1,6)
15    CONTINUE
      STOP
C
C
16    FORMAT (80A1)
17    FORMAT (I5)
18    FORMAT (8F10.0)
19    FORMAT (1H1,80A1,/,24X,31HLOG-PEARSON TYPE 3 DISTRIBUTION,/)
20    FORMAT (28X,26HMETHOD OF MOMENTS (DIRECT),/)
21    FORMAT (/,3X,43HMETHOD NOT APPLICABLE BECAUSE OF B VALUE OF,9X,E12
     1.5,/)
```

```
22    FORMAT (27X,28HMETHOD OF MOMENTS (INDIRECT),/)
23    FORMAT (9X,5HALPHA,5X,E12.5,14X,4HM1   ,6X,E12.5)
24    FORMAT (9X,5HBETA ,5X,E12.5,14X,4HM2   ,6X,E12.5)
25    FORMAT (9X,5HGAMMA,5X,E12.5,14X,4HSKEW,6X,E12.5,/)
26    FORMAT (3X,7HT,YEARS,4X,1H2,11X,1H5,10X,2H10,10X,2H20,10X,2H50,9X,
     13H100,/)
27    FORMAT (3X,1HX,3X,6E12.5,/,4X,1HT)
28    FORMAT (3X,1HS,3X,6E12.5,/,4X,1HT,/)
29    FORMAT (27X,28HMAXIMUM LIKELIHOOD PROCEDURE,/)
30    FORMAT (21X,5HTRIAL,11X,1HG,11X,4HF(G),/)
31    FORMAT (22X,I2,8X,E12.5,1X,E12.5)
32    FORMAT (/)
33    FORMAT (/,3X,50HSKEW IS NEGATIVE - DISTRIBUTION HAS AN UPPER BOUND
     1)/)
34    FORMAT (9X,5HL1   ,5X,E12.5,14X,4HM1   ,6X,E12.5)
35    FORMAT (9X,5HL2   ,5X,E12.5,14X,4HM2   ,6X,E12.5)
36    FORMAT (9X,5HL3   ,5X,E12.5,14X,4HSKEW,6X,E12.5,/)
      END
```

ST. MARYS RIVER AT STILLWATER - STATION NO. 01EO001
LOG-PEARSON TYPE 3 DISTRIBUTION

METHOD OF MOMENTS (DIRECT)

| | | | |
|---|---|---|---|
| L1 | .14555E+05 | M1 | .14555E+05 |
| L2 | .23870E+09 | M2 | .26865E+08 |
| L3 | .44403E+13 | SKEW | .13216E+01 |

| | | | |
|---|---|---|---|
| ALPHA | .25646E-01 | M1 | .95280E+01 |
| BETA | .17228E+03 | M2 | .11331E+00 |
| GAMMA | .51097E+01 | SKEW | .15237E+00 |

| T,YEARS | 2 | 5 | 10 | 20 | 50 | 100 |
|---|---|---|---|---|---|---|
| $X_T$ | .13623E+05 | .18188E+05 | .21258E+05 | .24243E+05 | .28186E+05 | .31217E+05 |
| $S_T$ | .64065E+03 | .96200E+03 | .13583E+04 | .19334E+04 | .29732E+04 | .39712E+04 |

METHOD OF MOMENTS (INDIRECT)

| | | | |
|---|---|---|---|
| ALPHA | .37788E-01 | M1 | .95285E+01 |
| BETA | .79726E+02 | M2 | .11384E+00 |
| GAMMA | .65158E+01 | SKEW | .22399E+00 |

| T,YEARS | 2 | 5 | 10 | 20 | 50 | 100 |
|---|---|---|---|---|---|---|
| $X_T$ | .13574E+05 | .18180E+05 | .21334E+05 | .24442E+05 | .28605E+05 | .31846E+05 |
| $S_T$ | .64099E+03 | .98097E+03 | .14078E+04 | .20317E+04 | .31684E+04 | .42692E+04 |

MAXIMUM LIKELIHOOD PROCEDURE

| TRIAL | G | F(G) |
|---|---|---|
| 1 | .86454E+01 | .14772E+01 |
| 2 | .85304E+01 | .66319E+00 |
| 3 | .83715E+01 | .28599E+00 |
| 4 | .81654E+01 | .11969E+00 |
| 5 | .79101E+01 | .48983E-01 |
| 6 | .76070E+01 | .19658E-01 |
| 7 | .72643E+01 | .77161E-02 |
| 8 | .69035E+01 | .29335E-02 |
| 9 | .65680E+01 | .10547E-02 |
| 10 | .63234E+01 | .33798E-03 |
| 11 | .62164E+01 | .82244E-04 |
| 12 | .62001E+01 | .97270E-05 |
| 13 | .61997E+01 | .19042E-06 |
| 14 | .61997E+01 | .92768E-10 |
| 15 | .61997E+01 | -.20009E-10 |

| | | | |
|---|---|---|---|
| ALPHA | .33645E-01 | M1 | .95285E+01 |
| BETA | .98939E+02 | M2 | .11199E+00 |
| GAMMA | .61997E+01 | SKEW | .20107E+00 |

| T,YEARS | 2 | 5 | 10 | 20 | 50 | 100 |
|---|---|---|---|---|---|---|
| $X_T$ | .13592E+05 | .18148E+05 | .21244E+05 | .24279E+05 | .28321E+05 | .31453E+05 |
| $S_T$ | .63209E+03 | .96537E+03 | .13754E+04 | .19638E+04 | .30232E+04 | .40411E+04 |

# CHAPTER 11

## TYPE III EXTREMAL DISTRIBUTION

### Introduction

The distribution discussed so far in Chapters 5-10 have been used primarily for frequency analysis of maximum events such as flood peaks. Two of these distributions (the 3-parameter log-normal and Pearson type III) have also been considered (Matalas (10)) suitable for analysis of minimum events such as low flows. The descriptions given for flood events remain valid for low flows except that the notation of probabilities must be changed.

The notation used so far for distributions of flood events has been

$$P(x) = 1 - \frac{m}{n+1} = 1 - 1/T \qquad (11\text{-}1)$$

where P(x) is the cumulative probability of an event being less than or equal to x and m is the order number of the recorded event, m being 1 for the maximum event and m being n for the minimum event. Following this notation, the larger the return period, T, the larger is the magnitude of the expected event. In the analysis of low flows, however, it is required that smaller events be associated with larger return periods and so a different notation must be used. If the recorded events are arranged in order of increasing magnitude with m being 1 for the minimum event and m being n for the maximum event then the cumulative probability of an event being less than or equal to x is given by

$$P(x) = \frac{m}{n+1} = \frac{1}{T} \qquad (11\text{-}2)$$

Another of the distributions considered by Matalas as suitable for low flow analysis is the type III extremal or Weibull distribution. Becuase this distribution has not been covered in any of the earlier chapters, a brief description follows. It has been noted by Condie (4) that if $\gamma = 0$ in the type III extremal distribution the lower limit is zero and the distribution might be suitable for flood analysis also.

The cumulative probability distribution is, (3):

$$P(x) = e^{-\left\{\frac{x-\gamma}{\beta-\gamma}\right\}^{\alpha}} \qquad (11\text{-}3)$$

and the probability density function is:

$$p(x) = \frac{\alpha}{\beta-\gamma}\{\frac{x-\gamma}{\beta-\gamma}\}^{\alpha-1} e^{-\{\frac{x-\gamma}{\beta-\gamma}\}^{\alpha}} \qquad (11\text{-}4)$$

where $\alpha$ is a scale parameter equal to the order of the lowest derivative of the probability function that is not zero at $x=\gamma$, $\beta$ is the characteristic drought (a location or central value parameter) and $\gamma$ is the lower limit to x.

Commonly, the transformation

$$y = \{\frac{x-\gamma}{\beta-\gamma}\}^{\alpha} \qquad (11\text{-}5)$$

is made (8) reducing the cumulative probability and probability density equations to

$$P(x) = e^{-y} \qquad (11\text{-}6)$$

and

$$p(x) = \frac{\alpha}{\beta-\gamma} y^{(\alpha-1)/\alpha} e^{-y} \qquad (11\text{-}7)$$

## Estimation of Paramaters

*Method of Moments* - The general expression for the rth moment, $\mu_r^{\gamma}$, about the lower bound, $\gamma$, of the type III extremal distribution is obtained from Equations 3-1 and 11-4 using $\gamma$ in place of the origin

$$\mu_r^{\gamma} = \int_0^{\infty} (x-\gamma)^r \frac{\alpha}{(\beta-\gamma)} \left\{\frac{x-\gamma}{\beta-\gamma}\right\}^{\alpha-1} \exp \left\{\frac{x-\gamma}{\beta-\gamma}\right\}^{\alpha} dx \qquad (11\text{-}8)$$

substituting y for $\{(x-\gamma)/(\beta-\gamma)\}^{\alpha}$ and simplifying gives

$$\mu_r^{\gamma} = (\beta-\gamma)^r \int_0^{\infty} y^{r/\alpha} e^{-y} dy \qquad (11\text{-}9)$$

$$\mu_r^{\gamma} = (\beta-\gamma)^r \Gamma(1+r/\alpha) \qquad (11\text{-}10)$$

where $\Gamma(x)$ is the gamma function equal to $(x-1)!$ for integer x. For non-integer values $\Gamma(x)$ can be computed from

$$\Gamma(x) = (x-1)(x-2)\ldots(x-r)\Gamma(y) \qquad (11\text{-}11)$$

for $x>1$, and for $x<1$ by

$$\Gamma(x) = \frac{\Gamma(y)}{x(x+1)(x+2)\ldots(x+r-1)} \qquad (11\text{-}12)$$

where $1\leq y\leq 2$ and $\Gamma(y)$ can be approximated by

$$\Gamma(y) = 1 + a_1(y-1) + a_2(y-1)^2 + a_3(y-1)^3 + a_4(y-1)^4 + a_5(y-1)^5 \tag{11-13}$$

with

$$a_1 = -0.5748646 \qquad a_4 = 0.4245549$$
$$a_2 = 0.9512363 \qquad a_5 = -0.1010678$$
$$a_3 = -0.6998588$$

A computer subroutine is available (9) which calculates $\Gamma(x)$ using similar relationships.

The moments about $\gamma$ may then be converted to origin moments or central moments as needed. As examples, for r=1

$$\mu_1^\gamma = (\beta-\gamma)\Gamma(1+1/\alpha) \tag{11-14}$$

and so the first moment about the origin, the mean, is

$$\mu_1' = \gamma + (\beta-\gamma)\Gamma(1+1/\alpha) \tag{11-15}$$

for r=2

$$\mu_2' = (\beta-\gamma)^2\Gamma(1+2/\alpha) \tag{11-16}$$

and from Equation 3-13

$$\mu_2 = \mu_2^\gamma - {\mu_1^\gamma}^2 \tag{11-17}$$

$$\mu_2 = \sigma^2 = (\beta-\gamma)^2 \{\Gamma(1+2/\alpha) - \Gamma^2(1+1/\alpha)\} \tag{11-18}$$

similarly

$$\mu_3 = (\beta-\gamma)^3 \{\Gamma(1+3/\alpha) - 3\Gamma(1+2/\alpha)\Gamma(1+1/\alpha) + 2\Gamma^3(1+1/\alpha)\} \tag{11-19}$$

$$\mu_4 = (\beta-\gamma)^4 \{\Gamma(1+4/\alpha) - 4\Gamma(1+3/\alpha)\Gamma(1+1/\alpha) + 6\Gamma(1+2/\alpha)\Gamma^2(1+1/\alpha) - 3\Gamma^4(1+1/\alpha)\} \tag{11-20}$$

$$\mu_5 = (\beta-\gamma)^5 \{\Gamma(1+5/\alpha) - 5\Gamma(1+4/\alpha)\Gamma(1+1/\alpha) + 10\Gamma(1+3/\alpha)\Gamma^2(1+1/\alpha) - 10\Gamma(1+2/\alpha)\Gamma^3(1+1/\alpha) + 4\Gamma^5(1+1/\alpha)\} \tag{11-21}$$

$$\mu_6 = (\beta-\gamma)^6 \{\Gamma(1+6/\alpha) - 6\Gamma(1+5/\alpha)\Gamma(1+1/\alpha) + 15\Gamma(1+4/\alpha)\Gamma^2(1+1/\alpha) - 20\Gamma(1+3/\alpha)\Gamma^3(1+1/\alpha) + 15\Gamma(1+2/\alpha)\Gamma^4(1+1/\alpha) - 5\Gamma^6(1+1/\alpha)\} \tag{11-22}$$

If two new variables are defined, $A_\alpha$ and $B_\alpha$, such that $A_\alpha$ is the standardized difference between the characteristic value and the mean and $B_\alpha$ is the standardized difference between the lower limit and the characteristic value,

$$A_\alpha = \frac{\beta-\mu}{\sigma} \qquad (11\text{-}23)$$

and

$$B_\alpha = \frac{\beta-\gamma}{\sigma} \qquad (11\text{-}24)$$

then, by substituting $\mu$ and $\sigma$ from Equations 11-15 and 11-18

$$B_\alpha = \{\Gamma(1+2/\alpha) - \Gamma^2(1+1/\alpha)\}^{-1/2} \qquad (11\text{-}25)$$

and

$$A_\alpha = \{1-\Gamma(1+1/\alpha)\}\ B_\alpha \qquad (11\text{-}26)$$

If the coefficient of skew, $\gamma_1$, is defined as usual

$$\gamma_1 = \frac{\mu_3}{\mu_2^{3/2}} \qquad (11\text{-}27)$$

then from Equations 11-18, 11-19 and 11-25

$$\gamma_1 = \{\Gamma(1+3/\alpha) - 3\Gamma(1+2/\alpha)\Gamma(1+1/\alpha) + 2\Gamma^3(1+1/\alpha)\}\ B_\alpha^3 \qquad (11\text{-}28)$$

an expression involving only functions of $\alpha$.

Thus if the sample coefficient of skew is computed as:

$$\hat{\gamma}_1 = \frac{n\ \Sigma\ (x-\bar{x})^3}{(n-2)[\Sigma(x-\bar{x})^2]^{3/2}} \qquad (11\text{-}29)$$

then $\alpha$ can be found by the solution of Equation 11-28. Knowing $\alpha$, the parameter $\beta$ can be obtained from Equation 11-23 as

$$\beta = \mu_1' + A_\alpha \sqrt{\mu_2} \qquad (11\text{-}30)$$

and $\gamma$ can be found from Equation 11-24 as

$$\gamma = \beta - B_\alpha \sqrt{\mu_2} \qquad (11\text{-}31)$$

To solve Equation 11-28 tables are available, (7), (2), relating $\alpha$ (generally as $1/\alpha$) to $\gamma_1$, $A_\alpha$ and $B_\alpha$. These tables are usually arranged in incremental steps of $1/\alpha$ so that for a computed sample skew a great deal of interpolation is needed to determine the corresponding values of $1/\alpha$, $A_\alpha$ and $B_\alpha$. In order to avoid this interpolation the following regression equation has been developed to enable $\alpha$ to be calculated directly from $\gamma_1$:

$$\alpha = 1/[a_1 + a_2\gamma_1 + a_3\gamma_1^2 + a_4\gamma_1^3 + a_5\gamma_1^4] \qquad (11\text{-}32)$$

$a_1 = 0.2777757913 \qquad a_4 = -0.0013038566$

$a_2 = 0.3132617714 \qquad a_5 = -0.0081523408$

$a_3 = 0.0575670910$

This polynomial is valid for a range of $\gamma_1$ from -1.02 to +2.00, has a multiple correlation coefficient of 0.9999 and a standard error of 0.0006575. Table 11-1 has been derived from this equation. For convenience Table 11-1 also gives values of the parameters, $A_\alpha$ and $B_\alpha$ as computed from Equations 11-26 and 11-25.

If the parameter $\gamma$ can be assumed to be zero, then a much simpler, if less accurate, method of estimating the remaining parameters, $\alpha$ and $\beta$, exists. If $x = \beta$ then, in Equation 11-3

$$P(\beta) = 1-e^{-1} = 0.632 \qquad (11\text{-}33)$$

The median of the type III extremal distribution, M, is obtained by substituting P = 0.50 in Equation 11-3,

$$M = \beta(\ln 2)^{1/\alpha} \qquad (11\text{-}34)$$

Table 11-1

Parameter $\alpha$ for Type III Extremal Distribution Tabulated as a Function of the Sample Coefficient of Skewness, $\gamma_1$

| $\gamma_1$ | $\alpha$ | $A_\alpha$ | $B_\alpha$ |
|---|---|---|---|
| -1.00 | 65.63043 | 0.44760 | 52.24465 |
| -0.90 | 26.26360 | 0.44229 | 21.47978 |
| -0.80 | 16.30207 | 0.43629 | 13.68443 |
| -0.70 | 11.73785 | 0.42952 | 10.10381 |
| -0.60 | 9.10978 | 0.42193 | 8.03409 |
| -0.50 | 7.39676 | 0.41343 | 6.67757 |
| -0.40 | 6.18962 | 0.40397 | 5.71462 |
| -0.30 | 5.29236 | 0.39350 | 4.99218 |
| -0.20 | 4.59923 | 0.38198 | 4.42770 |
| -0.10 | 4.04809 | 0.36938 | 3.97273 |
| 0.00 | 3.59997 | 0.35571 | 3.59692 |
| 0.10 | 3.22914 | 0.34098 | 3.28029 |
| 0.20 | 2.91791 | 0.32523 | 3.00911 |
| 0.30 | 2.65366 | 0.30851 | 2.77366 |
| 0.40 | 2.42717 | 0.29089 | 2.56682 |
| 0.50 | 2.23149 | 0.27246 | 2.38329 |
| 0.60 | 2.06133 | 0.25334 | 2.21910 |
| 0.70 | 1.91253 | 0.23367 | 2.07116 |
| 0.80 | 1.78181 | 0.21360 | 1.93718 |
| 0.90 | 1.66654 | 0.19329 | 1.81524 |
| 1.00 | 1.56457 | 0.17291 | 1.70391 |
| 1.10 | 1.47416 | 0.15265 | 1.60204 |
| 1.20 | 1.39386 | 0.13268 | 1.50873 |
| 1.30 | 1.32247 | 0.11318 | 1.42324 |
| 1.40 | 1.25900 | 0.09432 | 1.34501 |
| 1.50 | 1.20261 | 0.07626 | 1.27360 |
| 1.60 | 1.15260 | 0.05914 | 1.20866 |
| 1.70 | 1.10840 | 0.04311 | 1.14991 |
| 1.80 | 1.06954 | 0.02828 | 1.09714 |
| 1.90 | 1.03562 | 0.01477 | 1.05020 |
| 2.00 | 1.00634 | 0.00268 | 1.00900 |

If, therefore, a graph of original variable, x, versus cumulative probability is drawn, where cumulative probability is estimated from the sample using Equation 11-2, then values of $\beta$ and M can be extracted from the graph at P = 0.632 and P = 0.50. Substitution of $\beta$ and M in Equation 11-34 will then yield the sample estimate of $\alpha$.

*Maximum Likelihood* - Taking the logarithm of the product of the pdf (Equation 11-4) for $x_1 \rightarrow x_n$ the likelihood function is derived as:

$$\ln L = n \ln \alpha - n \ln (\beta-\gamma) + (\alpha-1) \sum_{i=1}^{n} \ln (x_i-\gamma)$$

$$- n (\alpha-1) \ln (\beta-\gamma) - (\beta-\gamma)^{-\alpha} \sum_{i=1}^{n} (x_i-\gamma)^{\alpha} \qquad (11\text{-}35)$$

Taking partial derivatives with respect to $\alpha$, $\beta$ and $\gamma$ yields:

$$\frac{\partial \ln L}{\partial \alpha} = \frac{n}{\alpha} + \sum_{i=1}^{n} \ln (x_i-\gamma) - n \ln (\beta-\gamma) - (\beta-\gamma)^{-\alpha} [ \sum_{i=1}^{n} (x_i-\gamma)^{\alpha} \cdot$$

$$\ln (x_i-\gamma) + \ln (\beta-\gamma) \ \Sigma \ (x_i-\gamma)^{\alpha}] \qquad (11\text{-}36)$$

$$\frac{\partial \ln L}{\partial \beta} = - \frac{n}{\beta-\gamma} - \frac{n(\alpha-1)}{\beta-\gamma} + \alpha(\beta-\gamma)^{-(\alpha+1)} \sum_{i=1}^{n} (x_i-\gamma)^{\alpha} \qquad (11\text{-}37)$$

$$\frac{\partial \ln L}{\partial \gamma} = \frac{n}{\beta-\gamma} - (\alpha-1) \sum_{i=1}^{n} (x_i-\gamma)^{-1} + \frac{n(\alpha-1)}{\beta-\gamma}$$

$$- \alpha(\beta-\gamma)^{-(\alpha+1)} \sum_{i=1}^{n} (x_i-\gamma)^{\alpha} + \alpha(\beta-\gamma)^{-\alpha} \sum_{i=1}^{n} (x_i-\gamma)^{\alpha-1} \qquad (11\text{-}38)$$

From 11-37 the expression

$$n(\beta-\gamma)^{\alpha} - \sum_{i=1}^{n} (x_i-\gamma)^{\alpha} = 0 \qquad (11\text{-}39)$$

is obtained. Using this as a substitution to eliminate $\beta$, Equations 11-36 and 11-38 reduce to

$$(\alpha- 1) \sum_{i=1}^{n} (x_i-\gamma)^{-1} - \frac{n\alpha \sum_{i=1}^{n} (x_i-\gamma)^{\alpha-1}}{\sum_{i=1}^{n} (x_i-\gamma)^{\alpha}} = 0 \qquad (11\text{-}40)$$

and

$$n + \alpha \sum_{i=1}^{n} \ln (x_i-\gamma) - \frac{n\alpha \sum_{i=1}^{n} (x_i-\gamma)^{\alpha} \ln (x_i-\gamma)}{\sum_{i=1}^{n} (x_i-\gamma)^{\alpha}} = 0 \qquad (11\text{-}41)$$

No further simplification is possible and Equations 11-40 and 11-41 must be solved as simultaneous equations. The procedure used is the computer program at the end of this chapter broadly follows the method of Condie and Nix (5). The interval between zero and the lowest recorded event is searched for the interval within which functions 11-40 and 11-41 cross. This is found by assuming successive values of $\gamma$ and calculating $\alpha_1$ from 11-40 and $\alpha_2$ from 11-41 both by Newton's method. Successive iterations then refine the interval containing the root until the required accuracy is found. Knowing $\alpha$ and $\gamma$, Equation 11-39 can be solved to give

$$\beta = \gamma + \left[\frac{\sum_{i=1}^{n} (x_i-\gamma)^{\alpha}}{n}\right]^{1/\alpha} \qquad (11\text{-}42)$$

*Other Methods* Various other methods of estimating parameters of the type III extremal distribution are available such as use of order statistics, smallest observed drought and least squares. Deininger and Westfield (6) compared the performance of several of these methods.

The method of smallest observed drought as described by Gumbel (7) leads to the three equations

$$\frac{\mu_1' - x_o}{\sqrt{\mu_2}} - \frac{(1-n^{-1/\alpha})\Gamma(1+1/\alpha)}{[\Gamma(1+2/\alpha) - \Gamma^2(1+1/\alpha)]^{1/2}} = 0 \qquad (11\text{-}43)$$

$$\gamma = x_o - \frac{\mu_1' - x_o}{(n^{1/\alpha}-1)} \qquad (11\text{-}44)$$

$$\beta = \frac{\mu_1' - \gamma}{\Gamma(1+1/\alpha)} + \gamma \qquad (11\text{-}45)$$

where $x_o$ is the smallest observed flow. Equation 11-43 can be solved for $\alpha$ by iteration using Newton's method and then $\gamma$ and $\beta$ can be found from 11-44 and 11-45. Gumbel (7) has also provided tables for the solution of 11-43.

## Frequency Factor

From Equation 11-5, 11-6, and 11-1 the following expressions are derived for the type III extremal distribution:

$$x = \gamma + y^{1/\alpha} (\beta-\gamma) \qquad (11\text{-}46)$$

and

$$y = -\ln (1-1/T) \qquad (11\text{-}47)$$

Values of y, the reduced variable, are given in Table 11-2 for some commonly used return periods.

Table 11-2

Values of the Reduced Variable, y, of the Type III Extremal Distribution for some Commonly Used Return Periods, T

| Return Period T | Reduced Variable y |
|---|---|
| 2 | 0.69315 |
| 5 | 0.22314 |
| 10 | 0.10536 |
| 20 | 0.05129 |
| 50 | 0.02020 |
| 100 | 0.01005 |

Knowing $\alpha$, $\beta$ and $\gamma$, and obtaining the reduced variable, y, from Table 11-2, the event magnitude, x, corresponding to return period, T, can be computed using Equation 11-46.

If Equations 11-46 and 11-47 are combined, then the following expression results:

$$x = \gamma + [-\ln (1-1/T)]^{1/\alpha} (\beta-\gamma) \qquad (11\text{-}48)$$

But from previous developments,

$$\beta-\gamma = B_\alpha \sigma \qquad (11\text{-}49)$$

and

$$\gamma = \mu - \sigma (B_\alpha - A_\alpha) \qquad (11\text{-}50)$$

where $\mu$ and $\sigma$ are the mean and standard deviation of the population event magnitudes as estimated from the sample, and $A_\alpha$ and $B_\alpha$ are as defined in Equations 11-26 and 11-25.

Substituting Equations 11-49 and 11-50 into Equation 11-48

$$x = \mu + \sigma \{(A_\alpha - B_\alpha) + B_\alpha [-\ln (1-1/T)]^{1/\alpha}\} \qquad (11\text{-}51)$$

Comparing Equation 11-51 with the standard frequency equation it is apparent that K, the frequency factor, is given by

$$K = A_\alpha + B_\alpha \{[-\ln (1-1/T)]^{1/\alpha} - 1\} \qquad (11\text{-}52)$$

This expression is dependent only upon the return period, T, and the coefficient of skew, $\gamma_1$, of the recorded events. Table 11-3 provides values of K for some typical values of T and $\gamma_1$.

Table 11-3

Frequency Factor for Use in Type III Extremal Distribution

| Coefficient of skew $\gamma_1$ | Cumulative Probability, P, % 50 | 80 | 90 | 95 | 98 | 99 |
|---|---|---|---|---|---|---|
| | Corresponding Return Period, T, Years 2 | 5 | 10 | 20 | 50 | 100 |
| -1.00 | 0.1567 | -0.7329 | -1.3134 | -1.8641 | -2.5680 | -3.0889 |
| -0.90 | 0.1446 | -0.7501 | -1.3215 | -1.8546 | -2.5232 | -3.0089 |
| -0.80 | 0.1321 | -0.7666 | -1.3282 | -1.8430 | -2.4766 | -2.9282 |
| -0.70 | 0.1189 | -0.7825 | -1.3332 | -1.8294 | -2.4280 | -2.8465 |
| -0.60 | 0.1051 | -0.7977 | -1.3366 | -1.8134 | -2.3771 | -2.7634 |
| -0.50 | 0.0906 | -0.8122 | -1.3382 | -1.7950 | -2.3239 | -2.6788 |
| -0.40 | 0.0754 | -0.8258 | -1.3379 | -1.7741 | -2.2683 | -2.5928 |
| -0.30 | 0.0595 | -0.8385 | -1.3356 | -1.7506 | -2.2103 | -2.5055 |
| -0.20 | 0.0428 | -0.8502 | -1.3313 | -1.7245 | -2.1502 | -2.4172 |
| -0.10 | 0.0255 | -0.8607 | -1.3248 | -1.6960 | -2.0881 | -2.3282 |
| 0.00 | 0.0075 | -0.8699 | -1.3161 | -1.6650 | -2.0244 | -2.2390 |
| 0.10 | -0.0110 | -0.8778 | -1.3053 | -1.6318 | -1.9595 | -2.1500 |
| 0.20 | -0.0300 | -0.8842 | -1.2923 | -1.5966 | -1.8938 | -2.0619 |
| 0.30 | -0.0493 | -0.8891 | -1.2773 | -1.5595 | -1.8277 | -1.9752 |
| 0.40 | -0.0689 | -0.8923 | -1.2603 | -1.5210 | -1.7616 | -1.8902 |
| 0.50 | -0.0885 | -0.8939 | -1.2415 | -1.4812 | -1.6961 | -1.8075 |
| 0.60 | -0.1081 | -0.8938 | -1.2209 | -1.4405 | -1.6315 | -1.7275 |
| 0.70 | -0.1275 | -0.8921 | -1.1989 | -1.3992 | -1.5682 | -1.6506 |
| 0.80 | -0.1466 | -0.8888 | -1.1757 | -1.3578 | -1.5068 | -1.5770 |
| 0.90 | -0.1651 | -0.8840 | -1.1515 | -1.3165 | -1.4473 | -1.5071 |
| 1.00 | -0.1829 | -0.8777 | -1.1266 | -1.2757 | -1.3903 | -1.4409 |
| 1.10 | -0.2000 | -0.8703 | -1.1013 | -1.2358 | -1.3359 | -1.3787 |
| 1.20 | -0.2162 | -0.8617 | -1.0758 | -1.1969 | -1.2842 | -1.3204 |
| 1.30 | -0.2313 | -0.8522 | -1.0505 | -1.1594 | -1.2356 | -1.2661 |
| 1.40 | -0.2454 | -0.8421 | -1.0255 | -1.1236 | -1.1901 | -1.2159 |
| 1.50 | -0.2583 | -0.8314 | -1.0013 | -1.0896 | -1.1477 | -1.1696 |
| 1.60 | -0.2701 | -0.8206 | -0.9780 | -1.0577 | -1.1086 | -1.1272 |
| 1.70 | -0.2807 | -0.8097 | -0.9558 | -1.0279 | -1.0728 | -1.0887 |
| 1.80 | -0.2900 | -0.7990 | -0.9351 | -1.0006 | -1.0403 | -1.0540 |
| 1.90 | -0.2983 | -0.7887 | -0.9159 | -0.9758 | -1.0112 | -1.0231 |
| 2.00 | -0.3053 | -0.7790 | -0.8985 | -0.9536 | -0.9854 | -0.9959 |

## Standard Error of Estimate

*Method of Moments* - The general equation for the standard error of estimate of a 3-parameter distribution has been given as

$$S_T^2 = \frac{\mu_2}{n}\left\{1 + K\gamma_1 + \frac{K^2}{4}\left[\gamma_2 - 1\right] + \frac{\partial K}{\partial \gamma_1}\left[2\gamma_2 - 3\gamma_1^2 - 6 + K\,(\gamma_3 - 6\gamma_1\gamma_2/4 - 10\gamma_1/4)\right] + \left(\frac{\partial K}{\partial \gamma_1}\right)^2\left[\gamma_4 - 3\gamma_3\gamma_1 - 6\gamma_2 + 9\gamma_1^2\gamma_2/4 + 35\gamma_1^2/4 + 9\right]\right\} \qquad (11\text{-}53)$$

where $\gamma_1$ is the coefficient of skew $\mu_3/\mu_2^{3/2}$; $\gamma_2$, the coefficient of kurtosis, $\mu_4/\mu_2^2$. $\gamma_3$ is $\mu_5/\mu_2^{5/2}$ and $\gamma_4$ is $\mu_6/\mu_2^3$ all obtainable from Equations 11-18 - 11-22. The frequency factor, K, can be substituted from Equation 11-52 leaving only $\partial K/\partial\gamma_1$ to be found. This may be done in two ways:

(a) analytically as

$$\frac{\partial K}{\partial\gamma_1} = \frac{\partial K}{\partial(1/\alpha)} \quad \frac{\partial(1/\alpha)}{\partial\gamma_1} \tag{11-54}$$

where $\partial K/\partial(1/\alpha)$ can be derived from Equation 11-52 and $\partial\gamma_1/\partial(1/\alpha)$ can be derived from Equation 11-28 noting that

$$\frac{\partial\ \Gamma(1+r/\alpha)}{\partial(1/\alpha)} = r\Gamma(1+r/\alpha).\psi(1+r/\alpha) \tag{11-55}$$

where $\psi$ is the psi or digamma function. This function can be evaluated by an asymptotic equation (1). Condie and Nix (5) found that to perserve accuracy at low values it was necessary to include a recurrence equation so that

$$\psi(z) \approx \ln\ (z+2) - \frac{1}{2(z+2)} - \frac{1}{12(z+2)^2} + \frac{1}{120(z+2)^4}$$

$$- \frac{1}{252(z+2)^6} - \frac{1}{(z+1)} - \frac{1}{z} \tag{11-56}$$

If y is - ln (1-1/T), $G_r$ is $\Gamma(1+r/\alpha)$ and $P_r$ is $\psi(1+r/\alpha)$ then the two partial derivatives are given by

$$\frac{\partial K}{\partial(1/\alpha)} = \frac{[\ln y.y^{1/\alpha} - G_1P_1] - [y^{1/\alpha} - G_1].[G_2 - G_1^2]^{-1}.[G_2P_2 - G_1^2P_1]}{[G_2 - G_1^2]^{1/2}} \tag{11-57}$$

and

$$\frac{\partial\gamma_1}{\partial(1/\alpha)} = 3\{[G_2 - G_1^2].[G_3P_3 - G_1G_2(P_1+2P_2) + 2G_1^3P_1]$$

$$- [G_3 - 3G_2G_1 + 2G_1^3].[G_2P_2 - G_1^2P_1]\}/[G_2 - G_1^2]^{5/2} \tag{11-58}$$

so that

$$\frac{\partial K}{\partial\gamma_1} = \{[\ln y.y^{1/\alpha} - G_1P_1] - [y^{1/\alpha} - G_1].[G_2 - G_1^2]^{-1}.[G_2P_2 - G_1^2P_1]\}.$$

$$[G_2 - G_1^2]^2/\{3([G_2 - G_1^2].[G_3P_3 - G_1G_2(P_1+2P_2)$$

$$+ 2G_1^3P_1] - [G_3 - 3G_2G_1 + 2G_1^3].[G_2P_2 - G_1^2P_1])\} \tag{11-59}$$

(b) numerically by computing $K_+$ and $K_-$ for $\gamma_{+,-} = \gamma_1 \pm \Delta\gamma$, from Equation 11-52. Then

$$\frac{\partial K}{\partial \gamma_1} \approx \frac{K_+ - K_-}{2\Delta\gamma_1} \qquad (11\text{-}60)$$

Using $\Delta\gamma_1 = 0.05$ this method gives good accuracy very simply.

Table 11-4 lists values of $\delta$ in

$$S_T = \delta\sqrt{\frac{\mu_2}{n}} \qquad (11\text{-}61)$$

for several values of return period, T, and coefficient of skew, $\gamma_1$, computed analytically.

Table 11-4

Values of Parameter $\delta$ for Use in Standard Error of Type III Extremal Distribution

| Coefficient of skew $\gamma_1$ | Cumulative Probability, P, % | | | | | |
|---|---|---|---|---|---|---|
| | 50 | 80 | 90 | 95 | 98 | 99 |
| | Corresponding Return Period, T, Years | | | | | |
| | 2 | 5 | 10 | 20 | 50 | 100 |
| - .8 | .9265 | 1.3665 | 1.8116 | 2.2267 | 2.6325 | 2.7650 |
| - .7 | .9743 | 1.3556 | 1.7517 | 2.1869 | 2.7877 | 3.2475 |
| - .6 | 1.0242 | 1.3492 | 1.6940 | 2.1413 | 2.8843 | 3.5450 |
| - .5 | 1.0710 | 1.3434 | 1.6356 | 2.0820 | 2.9084 | 3.6757 |
| - .4 | 1.0954 | 1.3259 | 1.5738 | 1.9846 | 2.7731 | 3.5067 |
| - .3 | 1.0886 | 1.2934 | 1.5063 | 1.8351 | 2.4456 | 3.0047 |
| - .2 | 1.0952 | 1.2624 | 1.4374 | 1.7132 | 2.2300 | 2.7011 |
| - .1 | 1.1065 | 1.2282 | 1.3709 | 1.6181 | 2.0938 | 2.5248 |
| .0 | 1.1157 | 1.1916 | 1.3042 | 1.5255 | 1.9631 | 2.3559 |
| .1 | 1.1244 | 1.1532 | 1.2374 | 1.4371 | 1.8437 | 2.2043 |
| .2 | 1.1318 | 1.1130 | 1.1711 | 1.3529 | 1.7336 | 2.0658 |
| .3 | 1.1394 | 1.0712 | 1.1078 | 1.2814 | 1.6496 | 1.9627 |
| .4 | 1.1460 | 1.0281 | 1.0467 | 1.2172 | 1.5775 | 1.8740 |
| .5 | 1.1517 | .9839 | .9905 | 1.1653 | 1.5236 | 1.8065 |
| .6 | 1.1567 | .9392 | .9414 | 1.1287 | 1.4905 | 1.7623 |
| .7 | 1.1605 | .8943 | .8981 | 1.1014 | 1.4661 | 1.7262 |
| .8 | 1.1636 | .8500 | .8646 | 1.0895 | 1.4583 | 1.7074 |
| .9 | 1.1657 | .8072 | .8422 | 1.0914 | 1.4630 | 1.7006 |
| 1.0 | 1.1671 | .7669 | .8319 | 1.1064 | 1.4788 | 1.7047 |
| 1.1 | 1.1678 | .7303 | .8348 | 1.1338 | 1.5049 | 1.7189 |
| 1.2 | 1.1681 | .6988 | .8507 | 1.1719 | 1.5394 | 1.7413 |
| 1.3 | 1.1680 | .6739 | .8792 | 1.2196 | 1.5815 | 1.7715 |
| 1.4 | 1.1676 | .6569 | .9186 | 1.2745 | 1.6291 | 1.8075 |
| 1.5 | 1.1669 | .6488 | .9673 | 1.3354 | 1.6816 | 1.8488 |
| 1.6 | 1.1658 | .6494 | 1.0218 | 1.3987 | 1.7355 | 1.8921 |
| 1.7 | 1.1643 | .6585 | 1.0807 | 1.4638 | 1.7908 | 1.9376 |
| 1.8 | 1.1622 | .6742 | 1.1406 | 1.5274 | 1.8446 | 1.9823 |
| 1.9 | 1.1596 | .6940 | 1.1987 | 1.5877 | 1.8952 | 2.0247 |
| 2.0 | 1.1564 | .7148 | 1.2523 | 1.6421 | 1.9405 | 2.0628 |

*Maximum Likelihood* - The variance of the T-year event, $S_T^2$, by maximum likelihood is

$$S_T^2 = \left(\frac{\partial x}{\partial \alpha}\right)^2 \text{var } \alpha + \left(\frac{\partial x}{\partial \beta}\right)^2 \text{var } \beta + \left(\frac{\partial x}{\partial \gamma}\right)^2 \text{var } \gamma$$

$$+ 2 \frac{\partial x}{\partial \alpha}\frac{\partial x}{\partial \beta} \text{cov } (\alpha,\beta) + 2 \frac{\partial x}{\partial \alpha}\frac{\partial x}{\partial \gamma} \text{cov } (\alpha,\gamma)$$

$$+ 2 \frac{\partial x}{\partial \beta}\frac{\partial x}{\partial \gamma} \text{cov } (\beta,\gamma) \qquad (11\text{-}62)$$

From the logarithm of the likelihood equation

$$\ln L = n \ln \alpha - n \ln (\beta-\gamma) + (\alpha-1) \sum_{i=1}^{n} \ln (x_i-\gamma)$$

$$- n(\alpha-1) \ln (\beta-\gamma) - (\beta-\gamma)^{-\alpha} \sum_{i=1}^{n} (x_i-\gamma)^{\alpha} \qquad (11\text{-}63)$$

the following derivatives are obtained:

$$\frac{\partial^2 \ln L}{\partial \alpha^2} = - \frac{n}{\alpha^2} - (\beta-\gamma)^{-\alpha} \left\{ \sum_{i=1}^{n} (x_i-\gamma)^{\alpha} \ln^2 (x_i-\gamma) \right.$$

$$\left. - \ln^2 (\beta-\gamma) \sum_{i=1}^{n} (x_i-\gamma)^{\alpha} \right\} \qquad (11\text{-}64)$$

$$\frac{\partial^2 \ln L}{\partial \alpha \partial \beta} = - \frac{n}{(\beta-\gamma)} + (\beta-\gamma)^{-(\alpha+1)} \{[\ln (\beta-\gamma) - 1] \sum_{i=1}^{n} (x_i-\gamma)^{\alpha}$$

$$+ \sum_{i=1}^{n} (x_i-\gamma)^{\alpha} \ln (x_i-\gamma)\} \qquad (11\text{-}65)$$

$$\frac{\partial^2 \ln L}{\partial \alpha \partial \gamma} = \frac{n}{(\beta-\gamma)} - \sum_{i=1}^{n} (x_i-\gamma)^{-1} + (\beta-\gamma)^{-\alpha} \left\{ \right.$$

$$\sum_{i=1}^{n} (x_i-\gamma)^{\alpha-1} [1 + \alpha \ln (x_i-\gamma)] + \alpha \ln (\beta-\gamma).$$

$$\left. \sum_{i=1}^{n} (x_i-\gamma)^{\alpha-1} \right\} + (\beta-\gamma)^{-(\alpha+1)} [1 - \alpha \ln (\beta-\gamma)].$$

$$\sum_{i=1}^{n} (x_i-\gamma)^{\alpha} - \alpha(\beta-\gamma)^{\alpha-1} \sum_{i=1}^{n} (x_i-\gamma)^{\alpha} \ln (x_i-\gamma) \qquad (11\text{-}66)$$

$$\frac{\partial^2 \ln L}{\partial \beta^2} = \frac{n\alpha}{(\beta-\gamma)^2} - \alpha(\alpha+1)(\beta-\gamma)^{-(\alpha+2)} \sum_{i=1}^{n} (x_i-\gamma)^{\alpha} \quad (11\text{-}67)$$

$$\frac{\partial^2 \ln L}{\partial \beta \partial \gamma} = - \frac{n\alpha}{(\beta-\gamma)^2} - \alpha^2(\beta-\gamma)^{-(\alpha+1)} \sum_{i=1}^{n} (x_i-\gamma)^{\alpha-1}$$

$$+ \alpha(\alpha+1)(\beta-\gamma)^{-(\alpha+2)} \sum_{i=1}^{n} (x_i-\gamma)^{\alpha} \quad (11\text{-}68)$$

$$\frac{\partial^2 \ln L}{\partial \gamma^2} = \frac{n\alpha}{(\beta-\gamma)^2} - (\alpha-1) \sum_{i=1}^{n} (x_i-\gamma)^{-2} - \alpha(\alpha+1)(\beta-\gamma)^{-(\alpha+2)}$$

$$\sum_{i=1}^{n} (x_i-\gamma)^{\alpha} + \alpha^2(\beta-\gamma)^{-(\alpha+1)} \sum_{i=1}^{n} (x_i-\gamma)^{\alpha-1}$$

$$- \alpha(\alpha-1)(\beta-\gamma)^{-\alpha} \sum_{i=1}^{n} (x_i-\gamma)^{\alpha-2} \quad (11\text{-}69)$$

now from Equation 11-46

$$\frac{\partial x}{\partial \alpha} = - \frac{1}{\alpha^2} (\beta-\gamma)\, y^{1/\alpha} \ln y \quad (11\text{-}70)$$

$$\frac{\partial x}{\partial \beta} = y^{1/\alpha} \quad (11\text{-}71)$$

and

$$\frac{\partial x}{\partial \gamma} = 1 - y^{1/\alpha} \quad (11\text{-}72)$$

where $y$ is $- \ln (1-1/T)$ and $T$ is the return period.

The information matrix is made up of the negative expectations of each of the above double derivatives. These expressions could be simplified, as has been done for other distributions, by finding analytical expressions for the expectations of

$$\sum_{i=1}^{n} (x_i-\gamma)^{\alpha} \quad (11\text{-}73)$$

$$\sum_{i=1}^{n} (x_i-\gamma)^{\alpha-1} \quad (11\text{-}74)$$

$$\sum_{i=1}^{n} [(x_i-\gamma)^{\alpha}, \ln (x_i-\gamma)] \quad (11\text{-}75)$$

and

$$\sum_{i=1}^{n} [(x_i-\gamma)^{\alpha}. \ln^2 (x_i-\gamma)] \quad (11\text{-}76)$$

so that, by inverting the matrix, analytic expressions for the variances and covariances needed in 11-62 could be found. For the type III extremal distribution this procedure has not been followed because of the complexity of expressions 11-64 - 11-69. Instead, the program at the end of this chapter evaluates each variances and covariances in terms of the matrix values.

Substitution of the numerical variances and covariances together with Equations 11-70 - 11-72 in Equation 11-62 yields the variance of the T-year event.

## Computer Program

The computer program which follows finds moment, maximum likelihood and smallest observed drought solutions to the type III extrėmal distribution. There is no maximum likelihood solution for the St. Mary's River at Stillwater and so a second data set, the South Thompson River at Chase, has been used to illustrate the results of this part of the program. The standard errors computed for the smallest observed drought solution have used the maximum likelihood method starting from computed $\alpha$, $\beta$ and $\gamma$. This may not be the minimum variance solution.

## References


1. Abramowitz, M. and I. A. Stegun, 1965, Handbook of Mathematical Functions, Dover Publications, New York, 1046 p.

2. Chin, W. O., 1967, Formulae and Tables for Computing and Plotting Drought Frequency Curves, Technical Bulletin No. 8, Inland Waters Branch, Ottawa.

3. Chow, V. T., 1964, Editor-in-Chief, Handbook of Applied Hydrology, McGraw-Hill.

4. Condie, R., 1976, personal communication.

5. Condie, R. and G. Nix, 1975, Modelling of Low Flow Frequency Distributions and Parameter Estimation, Int. Wat. Res. Assoc., Proc. Symp. on Water for Arid Lands, Tehran, Dec. 8-9.

6. Deininger, R. A. and J. D. Westfield, 1969, Estimation of the Parameters of Gumbel's Third Asymptotic Distribution by Different Methods, Wat. Res. Res., Vol. 5, No. 6, pp. 1238-1243.

7. Gumbel, E. J., 1958, Statistics of Extremes, Columbia University Press, 375 p.

8. Gumbel, E. J., 1966, Extreme Value Analysis of Hydrologic Data, Proc. Hydrology Symp. No. 5, NRC, Ottawa, pp. 149-169.

9. IBM, 1968, System/360 Scientific Subroutine Package (360A-CM-03X) Version III, Programmer's Manual, White Plains, New York, 454 p.

10. Matalas, N. C., 1963, Probability Distribution of Low Flows, USGS Professional Paper No. 434-A, 26 p.

```
      PROGRAM T3E(INPUT,OUTPUT,TAPE5=INPUT,TAPE6=OUTPUT)
C
C
C     COMPUTES METHOD OF MOMENTS AND MAXIMUM LIKELIHOOD ESTIMATES FOR
C     T YEAR EVENTS AND STANDARD ERRORS FOR TYPE 3 EXTREMAL
C     DISTRIBUTION
C     INPUT
C     TITLE
C     N NUMBER OF ANNUAL MAXIMUM EVENTS
C     X SERIES OF EVENTS
C
C
      REAL M1,M2,M3,K,IM,ML1,ML2,MLSKEW
      DIMENSION X(100), T(6), DUM(6), PS(3)
      DIMENSION XT(6), TITLE(80), ST(6), IM(9)
      DIMENSION AL1(25), AL2(25), DIF(25)
      T(1)=2
      T(2)=5
      T(3)=10
      T(4)=20
      T(5)=50
      T(6)=100
      A1=0.2777757913
      A2=0.3132617714
      A3=0.0575670910
      A4=-0.0013038566
      A5=-0.0081523408
      READ (5,30) TITLE
      READ (5,31) N
      XN=N
      READ (5,32) (X(I),I=1,N)
      WRITE (6,33) TITLE
      WRITE (6,34)
      A=0.0
      B=0.0
      C=0.0
      DO 1 I=1,N
      A=A+X(I)
      B=B+X(I)**2
      C=C+X(I)**3
1     CONTINUE
      M1=A/XN
      M2=(B/XN)-(A/XN)**2
      M3=(C/XN)+2.0*M1**3-3.0*M1*(B/XN)
      SKEW=M3/(M2**1.5)
      IF (SKEW.GT.2.00) WRITE (6,38)
      IF (SKEW.LT.-1.02) WRITE (6,38)
      AINV=A1+A2*SKEW+A3*SKEW**2+A4*SKEW**3+A5*SKEW**4
      ALPHA=1.0/AINV
      DO 2 JJ=1,6
      Z=1.0+JJ*AINV
      DUM(JJ)=0.0
2     CALL GMMMA (Z,DUM(JJ),IER)
      BA=1.0/((DUM(2)-DUM(1)**2)**0.5)
      AA=(1.0-DUM(1))*BA
      TEMP=DUM(2)-DUM(1)**2
      GA=SKEW
      GB=(DUM(4)-4.0*DUM(3)*DUM(1)+6.0*DUM(2)*DUM(1)**2-3.0*DUM(1)**4)/T
     1EMP**2
      GC=(DUM(5)-5.0*DUM(4)*DUM(1)+10.0*DUM(3)*DUM(1)**2-10.0*DUM(2)*DUM
     1(1)**3+4.0*DUM(1)**5)/TEMP**2.5
      GD=(DUM(6)-6.0*DUM(5)*DUM(1)+15.0*DUM(4)*DUM(1)**2-20.0*DUM(3)*DUM
     1(1)**3+15.0*DUM(2)*DUM(1)**4-5.0*DUM(1)**6)/TEMP**3
      DO 4 J=1,6
      Y=(-ALOG(1.0-1.0/T(J)))
      K=AA+BA*(Y**AINV-1.0)
      T1=1.0+K*GA+(GB-1.0)*(K**2)/4.0
      T2=2.0*GB-3.0*GA**2-6.0
      T3=K*(GC-1.5*GA*GB-2.5*GA)
      T4=GD-3.0*GC*GA-6.0*GB+2.25*GB*GA**2+8.75*GA**2+9.0
      DO 3 JJ=1,3
      Z=1.0+JJ*AINV
      DUM(JJ)=0.0
      CALL GMMMA (Z,DUM(JJ),IER)
      PS(JJ)=0.0
3     CALL PSI (Z,PS(JJ))
      G1=DUM(1)
      G2=DUM(2)
      G3=DUM(3)
      P1=PS(1)
```

```
      P2=PS(2)
      P3=PS(3)
      T5=(ALOG(Y)*Y**AINV)-G1*P1
      T6=(Y**AINV)-G1
      T7=G2-G1**2
      T8=G2*P2-P1*G1**2
      T9=G3*P3-G1*G2*(P1+2.0*P2)+2.0*P1*G1**3
      T10=G3-3.0*G2*G1+2.0*G1**3
      SLOPE=((T5-(T6/T7)*T8)*T7**2)/(3.0*(T7*T9-T10*T8))
      DELTA=T1+SLOPE*(T2+T3)+T4*SLOPE**2
      XT(J)=M1+K*SQRT(M2)
      ST(J)=SQRT(M2*DELTA/XN)
4     CONTINUE
      BETA=AA*SQRT(M2)+M1
      GAMMA=BETA-BA*SQRT(M2)
      WRITE (6,35) ALPHA,M1
      WRITE (6,36) BETA,M2
      WRITE (6,37) GAMMA,SKEW
      WRITE (6,39)
      WRITE (6,40) (XT(J),J=1,6)
      WRITE (6,41) (ST(J),J=1,6)
      WRITE (6,42)
      WRITE (6,47)
      AL=ALPHA
      GA=GAMMA
      XMIN=1000000.
      DO 5 I=1,N
5     IF (X(I).LT.XMIN) XMIN=X(I)
      DELTAX=XMIN/10.0
      GA=XMIN+DELTAX/2.0
      KCOUNT=0
6     KCOUNT=KCOUNT+1
      DO 13 IJK=1,20
      GA=GA-DELTAX
      ICOUNT=0
7     ICOUNT=ICOUNT+1
      A=0.0
      B=0.0
      C=0.0
      D=0.0
      F=0.0
      DO 8 I=1,N
      A=A+1.0/(X(I)-GA)
      B=B+(X(I)-GA)**AL
      C=C+((X(I)-GA)**(AL-1.0))
      D=D+((X(I)-GA)**(AL-1.0))*ALOG(X(I)-GA)
      F=F+((X(I)-GA)**AL)*ALOG(X(I)-GA)
8     CONTINUE
      FCN=((AL-1.0)/(XN*AL))*A-C/B
      FPN=(1.0/(XN*AL**2))*A-(B*D-C*F)/(B**2)
      BL=AL-FCN/FPN
      DELTA=ABS(0.0000001*BL)
      IF (ABS(BL-AL).LT.DELTA) GO TO 9
      IF (ICOUNT.GE.25) GO TO 21
      AL=BL
      GO TO 7
9     AL1(IJK)=AL
      JCOUNT=0
10    JCOUNT=JCOUNT+1
      A=0.0
      B=0.0
      C=0.0
      D=0.0
      DO 11 I=1,N
      A=A+((X(I)-GA)**AL)
      B=B+((X(I)-GA)**AL)*ALOG(X(I)-GA)
      C=C+((X(I)-GA)**AL)*(ALOG(X(I)-GA))**2
      D=D+ALOG(X(I)-GA)
11    CONTINUE
      FTN=XN+AL*D-XN*AL*B/A
      FUN=D-XN*AL*((A*C-B**2)/A**2)-XN*B/A
      BL=AL-FTN/FUN
      DELTA=ABS(0.0000001*BL)
      IF (ABS(BL-AL).LT.DELTA) GO TO 12
      IF (JCOUNT.GE.25) GO TO 21
      AL=BL
      GO TO 10
12    AL2(IJK)=AL
      DIF(IJK)=AL1(IJK)-AL2(IJK)
      IF (IJK.EQ.1) GO TO 13
```

```
      DELTA=ABS(0.0000001*AL1(IJK))
      IF (ABS(DIF(IJK)).LT.DELTA) GO TO 15
      IF (DIF(IJK-1).GE.0.0.AND.DIF(IJK).LT.0.0) GO TO 14
      IF (DIF(IJK-1).LT.0.0.AND.DIF(IJK).GT.0.0) GO TO 14
13    CONTINUE
      GO TO 21
14    GA=GA+DELTAX
      AL=(AL1(IJK)+AL1(IJK-1)+AL2(IJK)+AL2(IJK-1))/4.0
      WRITE (6,43) KCOUNT,AL,FTN
      IF (KCOUNT.GE.25) GO TO 21
      DELTAX=DELTAX/20.0
      GO TO 6
15    A=0.0
      B=0.0
      DO 16 I=1,N
      A=A+ALOG(X(I)-GA)
      B=B+(X(I)-GA)**AL
16    CONTINUE
      BE=GA+(B/XN)**(1.0/AL)
      ALPHA=AL
      GAMMA=GA
      BETA=BE
      AINV=1.0/ALPHA
      DO 17 JJ=1,3
      DUM(JJ)=0.0
      Z=1.0+JJ*AINV
17    CALL GMMMA (Z,DUM(JJ),IER)
      BA=1.0/((DUM(2)-DUM(1)**2)**0.5)
      AA=(1.0-DUM(1))*BA
      ML2=(BETA-GAMMA)/BA
      ML1=BETA-AA*ML2
      ML2=ML2**2
      MLSKEW=(DUM(3)-3.0*DUM(2)*DUM(1)+2.0*DUM(1)**3)*BA**3
      WRITE (6,44)
      WRITE (6,35) ALPHA,ML1
      WRITE (6,36) BETA,ML2
      WRITE (6,37) GAMMA,MLSKEW
      A=0.0
      B=0.0
      C=0.0
      D=0.0
      E=0.0
      F=0.0
      G=0.0
      H=0.0
      P=0.0
      Q=0.0
      R=0.0
      S=0.0
      U=0.0
      V=0.0
      W=0.0
      Z=0.0
      DO 18 I=1,N
      Z=X(I)-GAMMA
      A=A+(Z**ALPHA)*ALOG(Z)
      B=B+(Z**ALPHA)*ALOG(Z)*ALOG(Z)
      C=C+Z**ALPHA
      D=D+Z**(ALPHA-1.0)
      E=E+Z**(ALPHA-2.0)
      F=F+1.0/Z
      G=G+1.0/(Z**2)
      W=W+(Z**(ALPHA-1.0))*(1.0+ALPHA*ALOG(Z))
18    CONTINUE
      Z=BETA-GAMMA
      H=Z**(-ALPHA)
      P=Z**(-ALPHA-1.0)
      Q=Z**(-ALPHA-2.0)
      R=ALOG(Z)
      S=Z**(ALPHA-1.0)
      U=XN/Z
      V=XN/(Z**2)
      IM(1)=-XN/(ALPHA**2)-H*(B-(R**2)*C)
      IM(2)=-U+P*((R-1.0)*C+A)
      IM(3)=U-F+H*(W+ALPHA*R*D)+P*(1.0-ALPHA*R)*C-ALPHA*S*A
      TEMP1=ALPHA*(ALPHA+1.0)*Q*C
      TEMP3=ALPHA*(ALPHA-1.0)*H*E
      TEMP2=(ALPHA**2)*P*D
      IM(5)=ALPHA*V-TEMP1
      IM(6)=-ALPHA*V-TEMP2+TEMP1
```

```
      IM(9)=ALPHA*V-(ALPHA-1.0)*G-TEMP1+TEMP2-TEMP3
      IM(4)=IM(2)
      IM(7)=IM(3)
      IM(8)=IM(6)
      DO 19 KJ=1,9
19    IM(KJ)=-IM(KJ)
      DET=IM(1)*(IM(5)*IM(9)-IM(6)*IM(8))-IM(2)*(IM(4)*IM(9)-IM(6)*IM(7)
     1)+IM(3)*(IM(4)*IM(8)-IM(5)*IM(7))
      VARA=(IM(5)*IM(9)-IM(6)*IM(8))/DET
      VARB=(IM(1)*IM(9)-IM(3)*IM(7))/DET
      VARG=(IM(1)*IM(5)-IM(2)*IM(4))/DET
      COVAB=-(IM(4)*IM(9)-IM(6)*IM(7))/DET
      COVAG=(IM(4)*IM(8)-IM(5)*IM(7))/DET
      COVBG=-(IM(1)*IM(8)-IM(2)*IM(7))/DET
      DO 20 J=1,6
      Y=-ALOG(1.0/T(J))
      YA=Y**AINV
      YB=(-ALOG(1.0-1.0/T(J)))**AINV
      DXDA=-(BETA-GAMMA)*YA*ALOG(Y)/(ALPHA**2)
      DXDB=YA
      DXDG=1.0-YA
      XT(J)=GAMMA+(BETA-GAMMA)*YB
      ST(J)=SQRT(VARA*DXDA**2+VARB*DXDB**2+VARG*DXDG**2+2.0*DXDA*DXDB*CO
     1VAB+2.0*DXDA*DXDG*COVAG+2.0*DXDB*DXDG*COVBG)
20    CONTINUE
      WRITE (6,44)
      WRITE (6,39)
      WRITE (6,40) (XT(J),J=1,6)
      WRITE (6,41) (ST(J),J=1,6)
      GO TO 22
21    WRITE (6,45)
22    CONTINUE
      WRITE (6,46)
      WRITE (6,47)
      ASOD=ALPHA
      CONST=(M1-XMIN)/(SQRT(M2))
      ICOUNT=0
23    ICOUNT=ICOUNT+1
      F1=0.0
      F2=0.0
      P1=0.0
      P2=0.0
      T1=1.0+1.0/ASOD
      T2=1.0+2.0/ASOD
      CALL GMMMA (T1,F1,IER)
      CALL GMMMA (T2,F2,IER)
      CALL PSI (T1,P1)
      CALL PSI (T2,P2)
      F3=XN**(-1.0/ASOD)
      F4=ALOG(XN)
      F5=1.0/(ASOD**2)
      U=F1
      V=(F2-F1**2)**0.5
      W=F1*F3
      FCN=CONST-U/V+W/V
      DU=-F5*P1*F1
      DV=(1.0/V)*F5*(F1*P1-F2*P2)
      DW=F3*F1*F5*(F4-F1)
      FPN=(V*DW-W*DV)/(V**2)-(V*DU-U*DV)/(V**2)
      BSOD=ASOD-FCN/FPN
      WRITE (6,48) ICOUNT,BSOD,FCN
      DELTA=0.000001*BSOD
      IF (ABS(BSOD-ASOD).LT.DELTA) GO TO 24
      IF (ICOUNT.GE.25) GO TO 29
      ASOD=BSOD
      GO TO 23
24    ALPHA=BSOD
      AINV=1.0/ALPHA
      GAMMA=XMIN-(M1-XMIN)/(XN**AINV-1.0)
      BETA=((M1-GAMMA)/F1)+GAMMA
      DO 25 JJ=1,3
      DUM(JJ)=0.0
      Z=1.0+JJ*AINV
25    CALL GMMMA (Z,DUM(JJ),IER)
      BA=1.0/((DUM(2)-DUM(1)**2)**0.5)
      AA=(1.0-DUM(1))*BA
      M2=(BETA-GAMMA)/BA
      M1=BETA-AA*M2
      M2=M2**2
      SKEW=(DUM(3)-3.0*DUM(2)*DUM(1)+2.0*DUM(1)**3)*BA**3
```

```
      WRITE (6,44)
      WRITE (6,35) ALPHA,M1
      WRITE (6,36) BETA,M2
      WRITE (6,37) GAMMA,SKEW
      A=0.0
      B=0.0
      C=0.0
      D=0.0
      E=0.0
      F=0.0
      G=0.0
      H=0.0
      P=0.0
      Q=0.0
      R=0.0
      S=0.0
      U=0.0
      V=0.0
      W=0.0
      Z=0.0
      DO 26 I=1,N
      Z=X(I)-GAMMA
      A=A+(Z**ALPHA)*ALOG(Z)
      B=B+(Z**ALPHA)*ALOG(Z)*ALOG(Z)
      C=C+Z**ALPHA
      D=D+Z**(ALPHA-1.0)
      E=E+Z**(ALPHA-2.0)
      F=F+1.0/Z
      G=G+1.0/(Z**2)
      W=W+(Z**(ALPHA-1.0))*(1.0+ALPHA*ALOG(Z))
26    CONTINUE
      Z=BETA-GAMMA
      H=Z**(-ALPHA)
      P=Z**(-ALPHA-1.0)
      Q=Z**(-ALPHA-2.0)
      R=ALOG(Z)
      S=Z**(ALPHA-1.0)
      U=XN/Z
      V=XN/(Z**2)
      IM(1)=-XN/(ALPHA**2)-H*(B-(R**2)*C)
      IM(2)=-U+P*((R-1.0)*C+A)
      IM(3)=U-F+H*(W+ALPHA*R*D)+P*(1.0-ALPHA*R)*C-ALPHA*S*A
      TEMP1=ALPHA*(ALPHA+1.0)*Q*C
      TEMP3=ALPHA*(ALPHA-1.0)*H*E
      TEMP2=(ALPHA**2)*P*D
      IM(5)=ALPHA*V-TEMP1
      IM(6)=-ALPHA*V-TEMP2+TEMP1
      IM(9)=ALPHA*V-(ALPHA-1.0)*G-TEMP1+TEMP2-TEMP3
      IM(4)=IM(2)
      IM(7)=IM(3)
      IM(8)=IM(6)
      DO 27 KJ=1,9
27    IM(KJ)=-IM(KJ)
      DET=IM(1)*(IM(5)*IM(9)-IM(6)*IM(8))-IM(2)*(IM(4)*IM(9)-IM(6)*IM(7)
     1)+IM(3)*(IM(4)*IM(8)-IM(5)*IM(7))
      VARA=(IM(5)*IM(9)-IM(6)*IM(8))/DET
      VARB=(IM(1)*IM(9)-IM(3)*IM(7))/DET
      VARG=(IM(1)*IM(5)-IM(2)*IM(4))/DET
      COVAB=-(IM(4)*IM(9)-IM(6)*IM(7))/DET
      COVAG=(IM(4)*IM(8)-IM(5)*IM(7))/DET
      COVBG=-(IM(1)*IM(8)-IM(2)*IM(7))/DET
      DO 28 J=1,6
      Y=-ALOG(1.0/T(J))
      YA=Y**AINV
      YB=(-ALOG(1.0-1.0/T(J)))**AINV
      DXDA=-(BETA-GAMMA)*YA*ALOG(Y)/(ALPHA**2)
      DXDB=YA
      DXDG=1.0-YA
      XT(J)=GAMMA+(BETA-GAMMA)*YB
      ST(J)=SQRT(VARA*DXDA**2+VARB*DXDB**2+VARG*DXDG**2+2.0*DXDA*DXDB*CO
     1VAB+2.0*DXDA*DXDG*COVAG+2.0*DXDB*DXDG*COVBG)
28    CONTINUE
      WRITE (6,39)
      WRITE (6,40) (XT(J),J=1,6)
      WRITE (6,41) (ST(J),J=1,6)
29    CONTINUE
      STOP
C
30    FORMAT (80A1)
```

```
31    FORMAT (I5)
32    FORMAT (8F10.1)
33    FORMAT (1H1,80A1,/,26X,28HTYPE 3 EXTREMAL DISTRIBUTION,//)
34    FORMAT (33X,17HMETHOD OF MOMENTS,/)
35    FORMAT (9X,5HALPHA,5X,E12.5,14X,4HM1   ,6X,E12.5)
36    FORMAT (9X,5HBETA ,5X,E12.5,14X,4HM2   ,6X,E12.5)
37    FORMAT (9X,5HGAMMA,5X,E12.5,14X,4HSKEW,6X,E12.5,/)
38    FORMAT (/,42H WARNING- SKEW IS BEYOND PERMISSIBLE LIMIT,/)
39    FORMAT (3X,7HT,YEARS,4X,1H2,11X,1H5,10X,2H10,10X,2H20,10X,2H50,9X,
     13H100,/)
40    FORMAT (3X,1HX,3X,6E12.5,/,4X,1HT)
41    FORMAT (3X,1HS,3X,6E12.5,/,4X,1HT,/)
42    FORMAT (27X,28HMAXIMUM LIKELIHOOD PROCEDURE,/)
43    FORMAT (22X,I2,8X,E12.5,1X,E12.5)
44    FORMAT (/)
45    FORMAT (/,3X,45HMAXIMUM LIKELIHOOD DOES NOT CONVERGE IN ALPHA,/)
46    FORMAT (22X,35HMETHOD OF SMALLEST OBSERVED DROUGHT,/)
47    FORMAT (21X,5HTRIAL,12X,1HA,11X,4HF(A),/)
48    FORMAT (22X,I2,8X,E12.5,1X,E12.5)
      END

      SUBROUTINE GMMMA (XX,GX,IER)
      IF (XX-57.) 2,2,1
1     GX=0.39909-0.43429*(XX-1.0)+(XX-0.5)*ALOG10(XX-1.0)
      GX=10.0**GX
      RETURN
2     X=XX
      ERR=1.0E-6
      IER=0
      GX=1.0
      IF (X-2.0) 5,5,4
3     IF (X-2.0) 11,11,4
4     X=X-1.0
      GX=GX*X
      GO TO 3
5     IF (X-1.0) 6,12,11
6     IF (X-ERR) 7,7,10
7     Y=FLOAT(INT(X))-X
      IF (ABS(Y)-ERR) 13,13,8
8     IF (1.0-Y-ERR) 13,13,9
9     IF (X-1.0) 10,10,11
10    GX=GX/X
      X=X+1.0
      GO TO 9
11    Y=X-1.0
      GY=1.0+Y*(-0.5771017+Y*(+0.9858540+Y*(-0.8764218+Y*(+0.8328212+Y*(
     1-0.5684729+Y*(+0.2548205+Y*(-0.05149930)))))))
      GX=GX*GY
12    RETURN
13    IER=1
      RETURN
      END

      SUBROUTINE PSI (XX,YY)
      D=XX+2.0
      YY=ALOG(D)-(1.0/(2.0*D))-(1.0/(12.0*D**2))+(1.0/(120.0*D**4))-(1.0
     1/(252.0*D**6))-(1.0/(XX+1.0))-(1.0/XX)
      RETURN
      END
```

SOUTH THOMPSON RIVER AT CHASE - STATION NO. 08LE031
TYPE 3 EXTREMAL DISTRIBUTION

METHOD OF MOMENTS

| | | | |
|---|---|---|---|
| ALPHA | .24953E+01 | M1 | .25670E+04 |
| BETA | .27058E+04 | M2 | .21914E+06 |
| GAMMA | .14749E+04 | SKEW | .36831E+00 |

| T,YEARS | 2 | 5 | 10 | 20 | 50 | 100 |
|---|---|---|---|---|---|---|
| $X_T$ | .25377E+04 | .21497E+04 | .19744E+04 | .18492E+04 | .17326E+04 | .16697E+04 |
| $S_T$ | .78112E+02 | .71143E+02 | .72765E+02 | .84418E+02 | .10913E+03 | .12971E+03 |

MAXIMUM LIKELIHOOD PROCEDURE

| TRIAL | A | F(A) |
|---|---|---|
| 1 | .24747E+01 | -.49477E-09 |
| 2 | .27211E+01 | -.72760E-11 |
| 3 | .27205E+01 | -.15773E-06 |
| 4 | .27204E+01 | -.17462E-09 |

| | | | |
|---|---|---|---|
| ALPHA | .27204E+01 | M1 | .25650E+04 |
| BETA | .27131E+04 | M2 | .22392E+06 |
| GAMMA | .13722E+04 | SKEW | .26715E+00 |

| T,YEARS | 2 | 5 | 10 | 20 | 50 | 100 |
|---|---|---|---|---|---|---|
| $X_T$ | .25441E+04 | .21448E+04 | .19586E+04 | .18222E+04 | .16917E+04 | .16194E+04 |
| $S_T$ | .62835E+02 | .85641E+02 | .97692E+02 | .10761E+03 | .11871E+03 | .12604E+03 |

METHOD OF SMALLEST OBSERVED DROUGHT

| TRIAL | A | F(A) |
|---|---|---|
| 1 | .31037E+01 | .15732E+00 |
| 2 | .32370E+01 | .49611E-01 |
| 3 | .32833E+01 | .16694E-01 |
| 4 | .32994E+01 | .57480E-02 |
| 13 | .33080E+01 | .43358E-06 |

| | | | |
|---|---|---|---|
| ALPHA | .33080E+01 | M1 | .25670E+04 |
| BETA | .27283E+04 | M2 | .21914E+06 |
| GAMMA | .11609E+04 | SKEW | .75721E-01 |

| T,YEARS | 2 | 5 | 10 | 20 | 50 | 100 |
|---|---|---|---|---|---|---|
| $X_T$ | .25639E+04 | .21569E+04 | .19547E+04 | .17995E+04 | .16427E+04 | .15511E+04 |
| $S_T$ | .61339E+02 | .79128E+02 | .88175E+02 | .95476E+02 | .10350E+03 | .10873E+03 |

ST. MARYS RIVER AT STILLWATER - STATION NO. 01E0001
TYPE 3 EXTREMAL DISTRIBUTION

METHOD OF MOMENTS

| | | | |
|---|---|---|---|
| ALPHA | .12168E+01 | M1 | .75449E+02 |
| BETA | .80537E+02 | M2 | .39552E+04 |
| GAMMA | -.69849E+00 | SKEW | .14738E+01 |

| T,YEARS | 2 | 5 | 10 | 20 | 50 | 100 |
|---|---|---|---|---|---|---|
| $X_T$ | .59409E+02 | .22981E+02 | .12081E+02 | .63744E+01 | .25903E+01 | .11543E+01 |
| $S_T$ | .95556E+01 | .53226E+01 | .78101E+01 | .10801E+02 | .13654E+02 | .15047E+02 |

MAXIMUM LIKELIHOOD PROCEDURE

TRIAL A F(A)

MAXIMUM LIKELIHOOD DOES NOT CONVERGE IN ALPHA

METHOD OF SMALLEST OBSERVED DROUGHT

| TRIAL | A | F(A) |
|---|---|---|
| 1 | .11500E+01 | -.52949E-01 |
| 2 | .11518E+01 | .14625E-02 |
| 3 | .11518E+01 | -.51025E-05 |
| 4 | .11518E+01 | .21107E-07 |

| | | | |
|---|---|---|---|
| ALPHA | .11518E+01 | M1 | .75449E+02 |
| BETA | .79150E+02 | M2 | .39552E+04 |
| GAMMA | .32044E+01 | SKEW | .16183E+01 |

| T,YEARS | 2 | 5 | 10 | 20 | 50 | 100 |
|---|---|---|---|---|---|---|
| $X_T$ | .58450E+02 | .23855E+02 | .13968E+02 | .89660E+01 | .57702E+01 | .46038E+01 |
| $S_T$ | .62504E+01 | .12988E+02 | .17724E+02 | .22274E+02 | .28081E+02 | .32354E+02 |

# CHAPTER 12

## COMPARISON OF FREQUENCY DISTRIBUTIONS

### Introduction

Chapters 5-11 have described the use of various continuous probability distributions for estimating events at return periods larger than those of the recorded events. The question naturally arises as to which of these distributions to use for a particular data sample. The methods described in this chapter may be used to test the fit of a distribution to a particular sample. The larger question of whether or not the best fitting distribution is suitable for use in predicting event magnitudes outside the sample range will be discussed in Chapter 15. Tables 12-1 and 12-2 summarize the maximum likelihood event magnitudes and standard errors for several return periods computed for the annual maximum daily flows of the St. Mary's River at Stillwater, Nova Scotia, over the period 1916-1975 for the distributions of Chapters 5-11. In addition to the results taken from Chapters 5-11, calculations were made for the truncated normal distributions.

Table 12-1

Comparison of T-Year Event Magnitude Using Various Frequency Distributions, Thousands of Cubic Feet per Second

| | Cumulative Probability, P, % | | | | | |
|---|---|---|---|---|---|---|
| | 50 | 80 | 90 | 95 | 98 | 99 |
| | Corresponding Return Period, T, Years | | | | | |
| Frequency Distribution | 2 | 5 | 10 | 20 | 50 | 100 |
| Truncated Normal | 14.56 | 18.95 | 21.25 | 23.15 | 25.29 | 26.72 |
| 2-Parameter Lognormal | 13.69 | 18.36 | 21.41 | 24.29 | 28.01 | 30.80 |
| 3-Parameter Lognormal | 13.59 | 18.23 | 21.38 | 24.44 | 28.48 | 31.58 |
| Type I Extremal | 13.69 | 18.09 | 22.01 | 23.80 | 27.42 | 30.13 |
| Pearson Type III | 13.61 | 18.34 | 21.42 | 24.32 | 27.97 | 30.65 |
| Log-Pearson Type III | 13.59 | 18.15 | 21.24 | 24.27 | 28.32 | 31.45 |

Figure 12-1 shows the event magnitudes from each of these distributions plotted on type I extremal graph paper. The figure is for illustration only and normally many of these distributions would be plotted on other types of graph paper. The figure also shows those recorded events which have return periods greater than 2 years when computed from the equation $(n+1)/m$. It will be noted from the figure that the distributions are very closely grouped except for the truncated normal. The 2-parameter lognormal and the log-Pearson type III are so close as to be indistinguishable at this scale. Note also that all distributions predict a 100-year event considerably lower than the maximum event recorded in the 60-year sample data.

Table 12-2

Comparison of Standard Errors of T-Year Events Using Various Frequency Distributions, Thousands of Cubic Feet per Second

| | Cumulative Probability, P, % | | | | | |
|---|---|---|---|---|---|---|
| | 50 | 80 | 90 | 95 | 98 | 99 |
| | Corresponding Return Period, T, Years | | | | | |
| Frequency Distribution | 2 | 5 | 10 | 20 | 50 | 100 |
| Truncated Normal | .68 | .79 | .91 | 1.04 | 1.19 | 1.30 |
| 2-Parameter Lognormal | .62 | .96 | 1.30 | 1.68 | 2.22 | 2.67 |
| 3-Parameter Lognormal | .64 | .98 | 1.40 | 1.96 | 2.89 | 3.73 |
| Type I Extremal | .59 | .90 | 1.16 | 1.42 | 1.76 | 2.03 |
| Pearson Type III | .63 | .98 | 1.33 | 1.72 | 2.28 | 2.73 |
| Log-Pearson Type III | .63 | .96 | 1.37 | 1.96 | 3.02 | 4.04 |

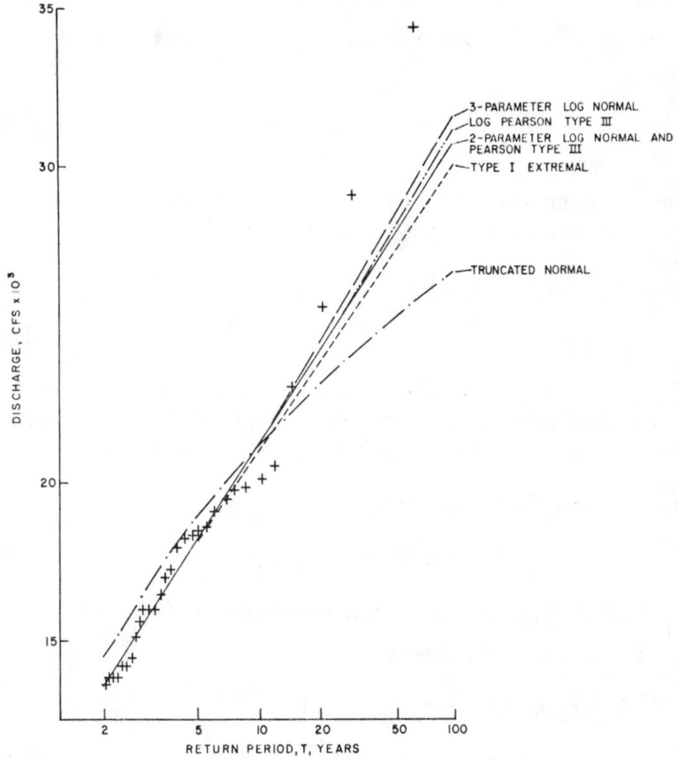


Figure 12-1. Comparison of frequency curves from various distributions - St. Mary's River at Stillwater, 1916-1975, Station No. 01E0001.

## Classical Tests

The two most commonly used tests of goodness of fit are the chi-square and Kolmogorov-Smirnov (Yevjevich (6)). An additional check on goodness of fit may be made by computing the sum of squares of differences between observed and computed event magnitudes.

*Chi-Square Test* - The statistic

$$\chi^2 = \sum_{j=i}^{k} \frac{(O_j - E_j)^2}{E_j} \tag{12-1}$$

is distributed assymptotically as chi-square with k-1 degrees of freedom where $O_j$ is the observed number of events in the $j_{th}$ class interval and $E_j$ is the number of events that would be expected from the theoretical distribution. If the class intervals are defined such that each interval corresponds to an equal probability then $E_j$ is n/k where n is the sample size and k is the number of class intervals and Equation 12-1 reduces to

$$\chi^2 = \frac{k}{n} \sum_{j=1}^{n} O_j^2 - n \tag{12-2}$$

according to Mann and Wald (4). The class intervals were computed for the various distributions as follows:

(a) Truncated Normal

$$CL = \bar{x} + tS \tag{12-3}$$

where $\bar{x}$ and S are the sample mean and standard deviation and t is the standard normal deviate corresponding to the probabilities of exceedence, P, listed in column 2 of Table 12-3.

(b) 2-Parameter Lognormal

$$CL = \exp\,(\bar{x}_n + tS_n) \tag{12-4}$$

where $\bar{x}_n$ and $S_n$ are the mean and standard deviation of the logarithms of the recorded events.

(c) 3-Parameter Lognormal

$$CL = a + \exp\,(\bar{x}_n + tS_{na}) \tag{12-5}$$

where a is the lower boundary of the distribution and $\bar{x}_{na}$ and $S_{na}$

Table 12-3

Comparison of Class Limits, Chi-Square and Kolmogorov-Smirnov Statistics and Standard Error for Different Distributions

St. Mary's River at Stillwater

| Class Interval | Probability | Truncated Normal | 2-Parameter Lognormal | 3-Parameter Lognormal | Type I Extremal | Pearson Type III | Log-Pearson Type III |
|---|---|---|---|---|---|---|---|
| 0 | 0 | 0 | 0 | 0 | 0 | 0 | 0 |
| 1 | 0.14286 | 9,100 | 9,666 | 9,589 | 9,136 | 9,780 | 9,681 |
| 2 | 0.28571 | 11,624 | 11,377 | 11,252 | 11,095 | 11,116 | 11,299 |
| 3 | 0.42857 | 13,618 | 12,939 | 12,784 | 12,836 | 12,518 | 12,810 |
| 4 | 0.57143 | 15,492 | 14,603 | 14,446 | 14,681 | 14,152 | 14,454 |
| 5 | 0.71429 | 17,486 | 16,609 | 16,526 | 16,947 | 16,264 | 16,484 |
| 6 | 0.85714 | 20,010 | 19,548 | 19,775 | 20,418 | 19,550 | 19,555 |
| 7 | 1.0 | ∞ | ∞ | ∞ | ∞ | ∞ | ∞ |
| Chi-Square | | 6.27 | 2.53 | 1.60 | 3.70 | 0.90 | 1.60 |
| $D_n$ | | 0.09 | 0.05 | 0.04 | 0.06 | 0.04 | 0.04 |
| SE | | 1,637 | 1,014 | 872 | 1,029 | 953 | 892 |

are the mean and standard deviation of the logarithms of the distribution x-a.

(d) Type I Extremal

$$CL = \bar{x} + [\frac{y_m - \mu}{\sigma}] S \qquad (12\text{-}6)$$

where $y_m$ is - ln (- ln P) and $\mu$ and $\sigma$ are the mean and standard deviation of the plotting positions.

(e) Pearson Type III

$$CL = \bar{x} + [\frac{\chi^2 \gamma_1}{4} - \frac{2}{\gamma_1}] S \qquad (12\text{-}7)$$

where $\chi^2$ is the value of chi-square at probability P and $8/\gamma_1^2$ degrees of freedom, $\gamma_1$ is the sample coefficient of skew.

(f) Log-Pearson Type III

$$CL = \exp\left(\bar{x}_n + \left[\frac{\chi^2 \gamma_n}{4} - \frac{2}{\gamma_n}\right] S_n\right) \qquad (12\text{-}8)$$

where $\gamma_n$ is the coefficient of skew of ln x.

The recorded events were then sorted in order of magnitude and the numbers of events within each class interval determined for each distribution.

Table 12-3 lists the computed class limits for each distribution together with the derived chi-square values. All the computed values of chi-square are less than the tabulated value at 95% and k-m-1 degrees of freedom, where m is the number of parameters estimated from the sample for the particular distribution. While no distinction can be made between the distributions on the basis of this test they may be listed in order of increasing magnitude of chi-square as:

Pearson type III
[3-parameter lognormal
log-Pearson type III]
2-parameter lognormal
type I extremal
truncated normal .

*Kolmogorov-Smirnov Test* - In order to avoid the loss of information due to grouping suffered by the chi-square test, other tests of goodness of fit have been developed such as the Neyman-Barton "smooth" tests, and the Cramer-von Mises $W^2$ test. The most important of these alternatives to chi-square is the Kolmogorov statistic, D, which is based on deviations of the sample distribution function P(x) from the completely specified continuous hypothetical distribution function $P_0(x)$, such that:

$$D_n = \max |P(x) - P_0(x)| \qquad (12\text{-}9)$$

Developed by Kolmogorov (3) in 1933, the test requires that the value of $D_n$ computed from the sample distribution be less than the tabulated value of $D_n$ at the required confidence level.

Table 12-3 shows, on the next to last line, the calculated value of $D_n$, the Kolmogorov-Smirnov statistic, for each distribution. In practice the values P(x) are obtained as $n_j/n$ where $n_j$ is the cumulative number of sample events at class limit j and n is the total number of events. $P_0(x)$ is then 1/k, 2/k, etc. where k is the number of class intervals. Given the 95% and 90% critical values of 0.18 and 0.16, for a sample size of 60, it can be seen that all distributions are well within the acceptance limits. On the basis of the Komogorov-Smirnov test the preferred order of distributions would be:

[3-parameter lognormal]
[Pearson type III]
[log-Pearson type III]
2-parameter lognormal
type I extremal
truncated normal .

*Least Squares* - A third method of comparing the fit of different distributions to a data sample is to compute the sum of squares of the differences between calculated and observed discharges. Bobée and Robitaille (1) used this method to compare calculated discharges at 2, 5, 10, 20, 50, and 100-year return periods with discharges at the same return periods interpolated (or extrapolated) from the recorded data. A difficulty with this method is that the interpolation itself is a form of distribution with which all the other distributions are then compared. Deininger and Westfield (2) compared the recorded events at their given return periods, (n+1)/m, to computed events from the distribution being tested at the same return periods (i.e. at 61 years for the maximum event in a 60-year record, at 30.5 years for the second largest event, etc.). The disadvantage of this procedure is the large number of computations needed.

The least squares method also has the disadvantage of being dependent on the plotting position equation used to compute the return periods of the observed events. Although this dependence will govern the absolute value of the sum of squares of differences for each distribution it will not affect the relative ranking of the distributions.

The program at the end of this chapter was used to compute the standard error of each of the six distributions as

$$SE_j = \left[\frac{\sum_{i=1}^{n} (x_i - y_i)^2}{n - m_j}\right]^{\frac{1}{2}} \qquad (12\text{-}10)$$

where $x_i$, i=1,...,n are the recorded events, $y_i$, i-1,...,n are the event magnitudes computed from the jth probability distribution at probabilities computed from the sorted ranks of $x_i$, i=1, ...,n, and $m_j$ is the number of parameters estimated for the jth distribution. The different distributions are represented in the program by subroutines abstracting the maximum likelihood procedure from each of the programs in Chapters 6-10 plus a simple procedure for the truncated normal distribution. The ranking of the distributions by order of least standard error is as follows:

3-parameter lognormal
log-Pearson type III
Pearson type III
2-parameter lognormal
type I extremal
truncated normal

The three goodness of fit tests are therefore in reasonable agreement that the best fitting of the distributions is the 3-parameter lognormal and the worst fitting is the truncated normal.

Bayesian Analysis

It has been suggested (5) that Bayesian analysis may provide a means of choosing one distribution from a set of distributions. By assuming a set of prior probabilities of choosing each distribution (all would be equal in the absence of an additional information) $p'(f_i)$, where $f_i$ represents the ith probability distribution function of the set, then the posterior probabilities can be derived by Bayesian analysis as

$$p''(f_i) = \frac{K_i}{K_*} \quad p'(f_i) \tag{12-11}$$

where $K_i$ is the marginal likelihood function of the observed data coming from the ith distribution and $K_*$ is a normalizing constant. By combining the probability distributions $f_i(q)$ and the posterior probabilities, $p''(f_i)$, as

$$f(q) = \Sigma \; p''(f_i) \; f_i(q) \tag{12-12}$$

the composite probability density function, $f(q)$, is the one which should be used in making inferences about future flood discharges. A Bayesian analyst would also require that $f_i(q)$ $i=1,\ldots,n$ all be Bayesian distributions so that $f(q)$ is a composite Bayesian distribution and accounts for both uncertainty in the choice of model and uncertainty in the parameter values of each model. As an example of this latter procedure, Wood and Rodriguez-Iturbe (5) have shown that using three Bayesian distributions, the normal, lognormal and exponential distributions, with equal prior probabilities of 1/3 the composite Bayesian distribution became

$$\hat{f}(q) = 0.996 \; f_1(q) + 0.004 \; f_2(q) \tag{12-13}$$

where $f_1(q)$ and $f_2(q)$ represent the exponential and lognormal pdf's and 0.996 and 0.004 are the posterior probabilities of these distributions. The posterior probability of the normal distribution was zero.

References


1. Bobée, B., and R. Robitaille, 1976, The Use of the Pearson Type 3 and log-Pearson Type 3 Distributions Revisited, submitted to Wat. Res. Res.

2. Deininger, R. A. and J. D. Westfield, 1969, Estimation of the Parameters of Gumbel's Third Asymptotic Distribution by Different Methods, Wat. Res. Res., Vol. 5, No. 6, pp. 1238-1243.

3. Kolmogorov, A., 1933, Sulla Determinazione Empirica di una Legge di Distribuzione, G. Ist. Ital. Attuari, 4, 83.

4. Mann, H. B. and A. Wald, 1942, On the Choice of the Number of Intervals in the Application of the Chi-Square Test, Ann. Math. Stat., 13, 306.

5. Wood, E. F. and I. Rodriguez-Iturbe, 1975, Bayesian Inference and Decision Making for Extreme Hydrologic Events, Wat. Res. Tes., Vol. 11, No. 4, pp. 533-542.

6. Yevjevich, V., 1972, Probability and Statistics in Hydrology, Water Resources Publications, Fort Collins, Colorado.

```
      PROGRAM SER(INPUT,OUTPUT,TAPE5=INPUT,TAPE6=OUTPUT)
C
C
C     PROGRAM TO COMPUTE THE STANDARD ERRORS OF EVENTS COMPUTED
C     FROM VARIOUS PROBABILITY DISTRIBUTIONS COMPARED TO THE
C     OBSERVED EVENT MAGNITUDES
C     INPUT
C     TITLE
C     N NUMBER OF ANNUAL MAXIMUM EVENTS
C     X SERIES OF EVENTS
C
C
      DIMENSION X(100), Y(100), Z(100), P(100), RP(100), T(100), TITLE(8
     10)
      REAL L1,L2,L3,LG
      REAL K,M1,M2,M3
      E=2.515517
      C1=0.802853
      C2=0.010328
      D1=1.432788
      D2=0.189269
      D3=0.001308
      READ (5,6) TITLE
      READ (5,7) N
      XN=N
      READ (5,8) (X(I),I=1,N)
      WRITE (6,9) TITLE
      A=0.0
      B=0.0
      C=0.0
      DO 1 I=1,N
      A=A+X(I)
      B=B+X(I)**2
      C=C+X(I)**3
1     CONTINUE
      M1=A/XN
      M2=(B/XN)-(A/XN)**2
      M3=(C/XN)+2.0*M1**3-3.0*M1*(B/XN)
      M2=M2*XN/(XN-1.0)
      G=M3/(M2**1.5)
      WRITE (6,11) M1
      WRITE (6,12) M2
      WRITE (6,13) G
      CALL SORTX (N,X)
      WRITE (6,10)
      WRITE (6,19) (X(I),I=1,N)
      WRITE (6,17)
      A=0.0
      B=0.0
      C=0.0
      DO 2 I=1,N
      Y(I)=ALOG(X(I))
      A=A+Y(I)
      B=B+Y(I)**2
      C=C+Y(I)**3
2     CONTINUE
      L1=A/XN
      L2=(B/XN)-(A/XN)**2
      L3=(C/XN)+2.0*L1**3-3.0*L1*(B/XN)
      L2=L2*XN/(XN-1.0)
      LG=L3/(L2**1.5)
      WRITE (6,14) L1
      WRITE (6,15) L2
      WRITE (6,16) LG
      DO 3 I=1,N
      XI=I
      P(I)=1.0-XI/(XN+1.0)
      RP(I)=1.0/(1.0-P(I))
      D=P(I)
      IF (D.GT.0.5) D=1.0-D
      W=SQRT(ALOG(1.0/(D**2)))
      T(I)=W-(E+C1*W+C2*W**2)/(1.0+D1*W+D2*W**2+D3*W**3)
      IF (P(I).LT.0.5) T(I)=-T(I)
3     CONTINUE
      WRITE (6,18)
      CALL IN (N,M1,M2,T,Z)
      WRITE (6,19) (Z(I),I=1,N)
      CALL SSQ (N,X,Z)
      WRITE (6,20)
      CALL IN (N,L1,L2,T,Z)
```

```
      DO 4 I=1,N
4     Z(I)=EXP(Z(I))
      WRITE (6,19) (Z(I),I=1,N)
      CALL SSQ (N,X,Z)
      WRITE (6,21)
      CALL LN3 (N,X,1,Z)
      WRITE (6,19) (Z(I),I=1,N)
      CALL SSQ (N,X,Z)
      WRITE (6,22)
      CALL TIE (N,X,M1,M2,RP,Z)
      WRITE (6,19) (Z(I),I=1,N)
      CALL SSQ (N,X,Z)
      WRITE (6,23)
      CALL PT3 (N,X,T,Z)
      WRITE (6,19) (Z(I),I=1,N)
      CALL SSQ (N,X,Z)
      WRITE (6,24)
      CALL PT3 (N,Y,T,Z)
      DO 5 I=1,N
5     Z(I)=EXP(Z(I))
      WRITE (6,19) (Z(I),I=1,N)
      CALL SSQ (N,X,Z)
      STOP
C
6     FORMAT (80A1)
7     FORMAT (I5)
8     FORMAT (8F10.0)
9     FORMAT (1H1,/,80A1,/)
10    FORMAT (3X,22HSORTED RECORDED EVENTS)
11    FORMAT (20X,9HMEAN OF X,16X,E12.5)
12    FORMAT (20X,13HVARIANCE OF X,12X,E12.5)
13    FORMAT (20X,9HSKEW OF X,16X,E12.5,/)
14    FORMAT (20X,15HMEAN OF LN(X)  ,10X,E12.5)
15    FORMAT (20X,19HVARIANCE OF LN(X)  ,6X,E12.5)
16    FORMAT (20X,15HSKEW OF LN(X)  ,10X,E12.5,/)
17    FORMAT (///)
18    FORMAT (3X,23HTRUNCATED NORMAL EVENTS)
19    FORMAT (3X,6E12.5)
20    FORMAT (3X,28H2 PARAMETER LOGNORMAL EVENTS)
21    FORMAT (3X,28H3 PARAMETER LOGNORMAL EVENTS)
22    FORMAT (3X,22HTYPE 1 EXTREMAL EVENTS)
23    FORMAT (3X,21HPEARSON TYPE 3 EVENTS)
24    FORMAT (3X,25HLOG-PEARSON TYPE 3 EVENTS)
      END
```

```
      SUBROUTINE SSQ (N,X,Z)
      DIMENSION X(1), Z(1)
      SUM=0.0
      DO 1 I=1,N
1     SUM=SUM+(Z(I)-X(I))**2
      XN=N
      SUM=SQRT(SUM/XN)
      WRITE (6,2) SUM
      RETURN
C
2     FORMAT (3X,17HSTANDARD ERROR IS,E12.5,//)
      END
```

```
      SUBROUTINE TN (N,XBAR,XVAR,T,X)
      DIMENSION X(1), T(1)
      XSTD=SQRT(XVAR)
      DO 1 I=1,N
1     X(I)=XBAR+T(I)*XSTD
      RETURN
      END
```

```
      SUBROUTINE SORTX (N,X)
C     SORTS IN DECREASING ORDER, X(1)=LARGEST
      DIMENSION X(1)
      K=N-1
      DO 2 L=1,K
      M=N-L
      DO 2 J=1,M
      IF (X(J)-X(J+1)) 1,1,2
1     XT=X(J)
      X(J)=X(J+1)
      X(J+1)=XT
2     CONTINUE
      RETURN
      END
```

```
      SUBROUTINE PT3 (N,X,SND,XT)
      DIMENSION X(1), SND(1), XT(1)
      XN=N
      ICOUNT=0
      XMIN=10000000.
      DO 1 I=1,N
1     IF (X(I).LT.XMIN) XMIN=X(I)
      GML=XMIN*0.99
2     ICOUNT=ICOUNT+1
      A=0.0
      B=0.0
      C=0.0
      R=0.0
      DO 3 I=1,N
      A=A+1.0/(X(I)-GML)
      B=B+(X(I)-GML)
      C=C+ALOG(X(I)-GML)
      R=R+1.0/((X(I)-GML)**2)
3     CONTINUE
      BETA=A/(A-(XN**2)/B)
      ALPHA=B/(XN*BETA)
      D=BETA+2.0
      PSI=ALOG(D)-(1.0/(2.0*D))-(1.0/(12.0*D**2))+(1.0/(120.0*D**4))-(1.
     10/(252.0*D**6))-(1.0/(BETA+1.0))-(1.0/BETA)
      FCN=-XN*PSI+C-XN*ALOG(ALPHA)
      TRI=(1.0/D)+(1.0/(2.0*D**2))+(1.0/(6.0*D**3))-(1.0/(30.0*D**5))+(1
     1.0/(42.0*D**7))-(1.0/(30.0*D**9))+(1.0/((BETA+1.0)**2))+(1.0/(BETA
     2**2))
      V=A-(XN**2)/B
      U=A
      W=(B/XN)-(XN/A)
      DU=R
      DV=R-(XN**3)/(B**2)
      DW=-1.0+(XN*R)/(A**2)
      FPN=-XN*TRI*((V*DU-U*DV)/(V**2))-A-XN*DW/W
      AS=GML-(FCN/FPN)
      DELTA=ABS(0.00000001*AS)
      IF (ABS(AS-GML).LT.DELTA) GO TO 4
      IF (ICOUNT.GT.25) GO TO 6
      GML=AS
      GO TO 2
4     CONTINUE
      GAMMA=AS
      DO 5 J=1,N
      T=SND(J)
      E=BETA**(1./3.)-1.0/(9.0*BETA**(2./3.))+T/(3.0*BETA**(1./6.))
      XT(J)=GAMMA+ALPHA*E**3
5     CONTINUE
6     CONTINUE
      RETURN
      END

      SUBROUTINE TIE (N,X,M1,M2,T,XT)
      DIMENSION X(1), T(1), XT(1)
      REAL M1,M2
      XN=N
      ICOUNT=0
      ALPHA=1.2825/(SQRT(M2))
      AML=ALPHA
1     ICOUNT=ICOUNT+1
      A=1.0/(AML**2)
      B=M1-1.0/AML
      C=0.0
      D=0.0
      E=0.0
      DO 2 I=1,N
      TEMP=EXP(-AML*X(I))
      C=C+TEMP
      D=D+TEMP*X(I)
      E=E+TEMP*X(I)**2
2     CONTINUE
      FCN=D-B*C
      FPN=B*D-E-A*C
      AS=AML-(FCN/FPN)
      DELTA=ABS(0.0000001*AS)
      IF (ABS(AS-AML).LT.DELTA) GO TO 3
      IF (ICOUNT.GT.25) GO TO 5
      AML=AS
      GO TO 1
3     CONTINUE
```

```
      ALPHA=AS
      BETA=(1.0/ALPHA)*ALOG(XN/C)
      DO 4 J=1,N
      YM=-ALOG(-ALOG(1.0-1.0/T(J)))
      XT(J)=BETA+YM/ALPHA
4     CONTINUE
5     CONTINUE
      RETURN
      END

      SUBROUTINE LN3 (N,X,SND,XT)
      DIMENSION X(1), SND(1), XT(1)
      REAL MU
      XN=N
      XMIN=10000000.
      DO 1 I=1,N
1     IF (X(I).LT.XMIN) XMIN=X(I)
      AML=XMIN*0.80
      ICOUNT=0
2     ICOUNT=ICOUNT+1
      A=0.0
      B=0.0
      C=0.0
      D=0.0
      E=0.0
      F=0.0
      P=0.0
      DO 3 I=1,N
      A=A+ALOG(X(I)-AML)
      B=B+(ALOG(X(I)-AML))**2
      P=P+(ALOG(X(I)-AML))**3
      C=C+1.0/((X(I)-AML))
      D=D+1.0/((X(I)-AML)**2)
      E=E+(1.0/((X(I)-AML)))*ALOG(X(I)-AML)
      F=F+(1.0/((X(I)-AML)**2))*ALOG(X(I)-AML)
3     CONTINUE
      G=(B/XN)-(A/XN)**2-(A/XN)
      H=(-2.0*E/XN)+(2.0*A/XN)*(C/XN)+(C/XN)
      FCN=C*G+E
      FPN=C*H+D*G+F-D
      AS=AML-(FCN/FPN)
      DELTA=ABS(0.0000001*AS)
      IF (ABS(AS-AML).LT.DELTA) GO TO 4
      IF (ICOUNT.GT.25) GO TO 6
      AML=AS
      GO TO 2
4     CONTINUE
      AML=AS
      MU=A/XN
      VAR=(B/XN)-(A/XN)**2
      VAR=VAR*XN/(XN-1)
      SKEW=(P/XN)+2.0*MU**3-3.0*MU*(B/XN)
      SD=SQRT(VAR)
      DO 5 J=1,N
      T=SND(J)
      Z=EXP(MU+T*SD)
      XT(J)=AML+Z
5     CONTINUE
6     CONTINUE
      RETURN
      END
```

ST. MARYS RIVER AT STILLWATER - STATION NO. 01F0001

| | |
|---|---|
| MEAN OF X | .14555E+05 |
| VARIANCE OF X | .27320E+08 |
| SKEW OF X | .12887E+01 |

SORTED RECORDED EVENTS

| | | | | | |
|---|---|---|---|---|---|
| .34400E+05 | .29100E+05 | .25600E+05 | .23000E+05 | .20600E+05 | .20100E+05 |
| .19900E+05 | .19900E+05 | .19500E+05 | .19200E+05 | .18600E+05 | .18500E+05 |
| .18300E+05 | .18200E+05 | .18000E+05 | .17200E+05 | .16900E+05 | .16400E+05 |
| .16100E+05 | .16100E+05 | .16000E+05 | .15600E+05 | .15100E+05 | .14500E+05 |
| .14300E+05 | .14300E+05 | .13900E+05 | .13900E+05 | .13900E+05 | .13600E+05 |
| .13100E+05 | .13100E+05 | .13000E+05 | .13000E+05 | .12900E+05 | .12700E+05 |
| .12400E+05 | .12300E+05 | .12300E+05 | .12200E+05 | .11900E+05 | .11900E+05 |
| .11800E+05 | .11800E+05 | .11600E+05 | .11000E+05 | .10700E+05 | .10400E+05 |
| .10300E+05 | .10200E+05 | .10200E+05 | .99000E+04 | .90200E+04 | .83900E+04 |
| .82100E+04 | .82100E+04 | .81800E+04 | .80400E+04 | .71300E+04 | .67000E+04 |

```
                    MEAN OF LN(X)                  .95285E+01
                    VARIANCE OF LN(X)              .11384E+00
                    SKEW OF LN(X)                  .18649E+00

TRUNCATED NORMAL EVENTS
   .25715E+05  .24181E+05  .23196E+05  .22447E+05  .21832E+05  .21303E+05
   .20836E+05  .20414E+05  .20027E+05  .19668E+05  .19332E+05  .19014E+05
   .18712E+05  .18424E+05  .18146E+05  .17879E+05  .17620E+05  .17368E+05
   .17122E+05  .16883E+05  .16648E+05  .16417E+05  .16190E+05  .15966E+05
   .15744E+05  .15525E+05  .15307E+05  .15091E+05  .14876E+05  .14662E+05
   .14448E+05  .14233E+05  .14018E+05  .13802E+05  .13584E+05  .13365E+05
   .13144E+05  .12920E+05  .12693E+05  .12462E+05  .12227E+05  .11987E+05
   .11742E+05  .11490E+05  .11231E+05  .10963E+05  .10686E+05  .10397E+05
   .10095E+05  .97772E+04  .94409E+04  .90819E+04  .86951E+04  .82733E+04
   .78061E+04  .72778E+04  .66626E+04  .59135E+04  .49281E+04  .33946E+04
STANDARD ERROR IS  .16373E+04

2 PARAMETER LOGNORMAL EVENTS
   .28251E+05  .25588E+05  .24011E+05  .22878E+05  .21987E+05  .21250E+05
   .20619E+05  .20065E+05  .19570E+05  .19122E+05  .18711E+05  .18331E+05
   .17977E+05  .17645E+05  .17332E+05  .17035E+05  .16753E+05  .16483E+05
   .16224E+05  .15974E+05  .15734E+05  .15501E+05  .15276E+05  .15056E+05
   .14843E+05  .14634E+05  .14430E+05  .14230E+05  .14034E+05  .13841E+05
   .13651E+05  .13463E+05  .13278E+05  .13094E+05  .12911E+05  .12730E+05
   .12549E+05  .12369E+05  .12189E+05  .12009E+05  .11828E+05  .11646E+05
   .11463E+05  .11278E+05  .11091E+05  .10901E+05  .10708E+05  .10510E+05
   .10307E+05  .10098E+05  .98810E+04  .96547E+04  .94166E+04  .91636E+04
   .88914E+04  .85933E+04  .82587E+04  .78689E+04  .73839E+04  .66880E+04
STANDARD ERROR IS  .10144E+04

3 PARAMETER LOGNORMAL EVENTS
   .29367E+05  .26300E+05  .24516E+05  .23248E+05  .22262E+05  .21452E+05
   .20763E+05  .20162E+05  .19629E+05  .19148E+05  .18710E+05  .18306E+05
   .17931E+05  .17582E+05  .17253E+05  .16943E+05  .16648E+05  .16368E+05
   .16100E+05  .15843E+05  .15596E+05  .15357E+05  .15127E+05  .14904E+05
   .14687E+05  .14476E+05  .14270E+05  .14069E+05  .13872E+05  .13680E+05
   .13490E+05  .13304E+05  .13120E+05  .12938E+05  .12759E+05  .12580E+05
   .12404E+05  .12228E+05  .12053E+05  .11878E+05  .11703E+05  .11528E+05
   .11353E+05  .11176E+05  .10998E+05  .10817E+05  .10634E+05  .10448E+05
   .10257E+05  .10062E+05  .98601E+04  .96506E+04  .94314E+04  .91998E+04
   .89520E+04  .86826E+04  .83826E+04  .80363E+04  .76108E+04  .70106E+04
STANDARD ERROR IS  .87215E+03

TYPE 1 EXTREMAL EVENTS
   .29197E+05  .25474E+05  .23867E+05  .22717E+05  .21817E+05  .21075E+05
   .20441E+05  .19887E+05  .19394E+05  .18949E+05  .18541E+05  .18166E+05
   .17816E+05  .17490E+05  .17182E+05  .16891E+05  .16614E+05  .16350E+05
   .16097E+05  .15855E+05  .15620E+05  .15394E+05  .15175E+05  .14962E+05
   .14755E+05  .14553E+05  .14356E+05  .14162E+05  .13973E+05  .13786E+05
   .13603E+05  .13421E+05  .13242E+05  .13065E+05  .12890E+05  .12715E+05
   .12541E+05  .12368E+05  .12195E+05  .12022E+05  .11848E+05  .11674E+05
   .11498E+05  .11320E+05  .11140E+05  .10957E+05  .10771E+05  .10580E+05
   .10384E+05  .10182E+05  .99718E+04  .97519E+04  .95201E+04  .92730E+04
   .90060E+04  .87122E+04  .83805E+04  .79908E+04  .75006E+04  .67838E+04
STANDARD ERROR IS  .10292E+04

PEARSON TYPE 3 EVENTS
   .29745E+05  .26020E+05  .24386E+05  .23204E+05  .22271E+05  .21497E+05
   .20832E+05  .20249E+05  .19726E+05  .19253E+05  .18819E+05  .18418E+05
   .18044E+05  .17693E+05  .17362E+05  .17049E+05  .16751E+05  .16467E+05
   .16194E+05  .15932E+05  .15679E+05  .15435E+05  .15199E+05  .14969E+05
   .14746E+05  .14529E+05  .14317E+05  .14109E+05  .13906E+05  .13706E+05
   .13510E+05  .13317E+05  .13126E+05  .12938E+05  .12752E+05  .12567E+05
   .12384E+05  .12202E+05  .12021E+05  .11840E+05  .11659E+05  .11478E+05
   .11297E+05  .11115E+05  .10932E+05  .10747E+05  .10560E+05  .10370E+05
   .10176E+05  .99779E+04  .97745E+04  .95644E+04  .93459E+04  .91169E+04
   .88741E+04  .86132E+04  .83270E+04  .80032E+04  .76170E+04  .71001E+04
STANDARD ERROR IS  .95289E+03

LOG-PEARSON TYPE 3 EVENTS
   .29216E+05  .26135E+05  .24354E+05  .23096E+05  .22119E+05  .21318E+05
   .20639E+05  .20047E+05  .19522E+05  .19049E+05  .18619E+05  .18222E+05
   .17855E+05  .17512E+05  .17189E+05  .16885E+05  .16596E+05  .16322E+05
   .16059E+05  .15807E+05  .15565E+05  .15331E+05  .15106E+05  .14887E+05
   .14674E+05  .14467E+05  .14265E+05  .14068E+05  .13875E+05  .13686E+05
   .13500E+05  .13317E+05  .13136E+05  .12957E+05  .12781E+05  .12605E+05
   .12431E+05  .12258E+05  .12085E+05  .11913E+05  .11740E+05  .11567E+05
   .11393E+05  .11218E+05  .11042E+05  .10863E+05  .10681E+05  .10495E+05
   .10306E+05  .10111E+05  .99090E+04  .96992E+04  .94792E+04  .92462E+04
   .89962E+04  .87236E+04  .84188E+04  .80651E+04  .76276E+04  .70037E+04
STANDARD ERROR IS  .89223E+03
```

## CHAPTER 13

## REGIONAL ANALYSIS

### Introduction

Chapters 5-11 have described some of the probability distributions that can be used to carry out a frequency analysis on a set of observed or computed data. Using any one of those techniques the event magnitude corresponding to a given probability of occurrency can be determined. This event magnitude will, however, apply only to the exact location at which the original observations were made. Frequently in hydrology it is necessary to estimate event magnitudes at sites where no observations have been taken. As an example, the design of a highway culvert may require the estimation of a design flood for a small ungauged catchment area. Regional analysis is the term given to techniques which make this estimation possible.

In addition, as noted in Chapter 1, the use of more than one set of data tends to reduce the sampling error and, even for a gauged site, will produce more reliable event estimates than a single station frequency analysis. Indeed, as discussed later, the Bayesian approach to frequency analysis might be to use distributions of parameters obtained from a regional analysis in conjunction with the point values to account for the uncertainty in the estimates.

The earliest approach to the regionalization problem was to use empirical equations relating floodflow, Q, to drainage area, A, within a particular region (3) such as:

$$Q = cA^n \qquad (13\text{-}1)$$

where c and n are constants. Other types of empirical equation (some still in use, such as the Rational Formula) related floodflow to rainfall intensity and area, as:

$$Q = ciA \qquad (13\text{-}2)$$

where c is a runoff coefficient, i is the rainfall intensity and A is the area. The objective of all these equations was to extrapolate from gauged basins to ungauged basins by means of parameters which could be estimated (rainfall intensity) or measured from maps (area).

Other methods in use include that designed by Coulson (10) for Southern Ontario. From the records of 59 gauging stations in the area, together with an isohyetal map of mean annual precipitation, a map of lines of equal mean annual runoff was drawn. Rewriting the standard frequency equation (see Chapter 3) as

$$Q_T = \bar{Q}\ (K\ z + 1) \qquad (13\text{-}3)$$

where $Q_T$ is the event magnitude at the required return period, T, $\bar{Q}$ is the mean annual runoff, K is a frequency factor depending on the probability distribution used and z is the coefficient of variation. For an ungauged drainage basin for which an estimate of $Q_T$ is required, Coulson obtained $\bar{Q}$ by planimetering the mean annual runoff isoline map for the particular basin. Using a Pearson type III distribution the value of K, the frequency factor, depends upon the return period, T, and the coefficient of skew of the distribution, $\gamma_1$. It was found in Southern Ontario that the coefficient of skew could best be derived from the coefficient of variation as:

$$\gamma_1 = 2z \qquad (13\text{-}4)$$

while the coefficient of variation could be obtained from the drainage area, A, as (14):

$$z = 0.35 - 0.03 \log (A + 1) \qquad (13\text{-}5)$$

It should be noted that this method was used only for annual flows and not for instantaneous maxima or minima.

The USGS currently uses a method described as an "index-flood" technique (12). There are two major parts to this method. Firstly, basic dimensionless frequency curves are drawn representing the ratios of the floods at various frequencies to the mean annual flood for each gauged basin. Secondly, relationships are developed between the characteristics of drainage areas and the mean annual flood. Combining the mean annual flood with a regional frequency curve enables flood magnitudes to be estimated at any location within the region.

Many modifications of the original index-flood method have been made (3, 11), mainly to try and increase the number of independent variables used to transfer the hydrologic information. In general there are two types of variables used: (a) physiographic characteristics such as drainage area, elevation, slope, percent of basin covered by lakes, swamps, etc. and (b) hydrometeorologic variables such as mean annual precipitation and mean annual temperature.

One of the major developments since the index-flood technique is the "square-grid" method (26). Originally programmed for mean annual runoff this method has been adapted for frequency analyses, simulation and modelling (16, 25, 18).

Alternate methods include the use of standard single-station frequency distributions modified for use as regional distributions and the regional record maxima technique.

Many of the methods used in regional analysis depend upon inter-station correlation of streamflows in order to produce time series with a uniform period of record. The final section of this chapter discusses information transfer together with the concept of effective number of years of record and number of stations.

## Index-Flood Method

The basic idea behind the index-flood method (12) is to increase the reliability of the frequency characteristics within a region. If, within a hydrologically homogeneous area, a number of hydrometric stations have been operating and recording the effects of the same meteorologic factors then a combination of these records will provide, not a longer record, but a more reliable record. The following brief description of the index-flood method does not include all the computational details of the procedure. These can be obtained, if required, from Dalrymple (12).

Firstly, the data sets available within a region are listed, unsuitable stations eliminated, and a common period of record selected. Generally stations having less than 5 years of record of gauging and all regulated or controlled streams are excluded. Since streamflows are not pure random but contain trends and periodicities the period over which measurements are made becomes important when records are combined. A bar graph showing the period of record of each gauge is useful in determining which base period to use. The base period should be planned so as to include the maximum information content i.e. maximum number of station-years. Missing data points may be filled in by inter-station correlations (see section on regional record maxima). Data points filled in in this way are not used directly but only as aids in assigning representative return periods to the recorded events.

The index-flood method next computes return periods, T, for each recorded event for each station in the region using the equation:

$$T = \frac{n + 1}{m} \qquad (13\text{-}6)$$

where n is the sample size and m is the order number of an event; m = 1 for the maximum event and m = n for the minimum event. For each station a graph of T versus event magnitude is plotted and a smooth curve drawn through the points. No attempt is made to force a straight line fit or to fit any mathematical distribution. The mean annual event for the station is then picked off the smooth curve at the point T = 2.33. This is a theoretical result taken from the type I extremal distribution (see Chapter 8). Benson (2) has confirmed experimentally that the mean annual event (i.e. the mean of all the observed annual maxima) does occur with a return period of 2.33 years. It is preferred in the index-flood method to derive the mean annual event graphically rather than arithmetically.

Dalrymple (12) has described a test which should be used at this stage of the index-flood procedure to check for regional hydrologic homogeneity. If the standard error of estimate of the reduced variable, y, in a type I extremal distribution is given by:

$$\sigma_y = \frac{e^y}{\sqrt{n}} \sqrt{\frac{1}{T-1}} \qquad (13\text{-}7)$$

then, assuming a normal distribution of the estimates, 95% of the estimates will lie within $\pm 2\sigma_y$ of the most probable value. If T, the return period of the estimate, is taken as 10 years, then

$$2\sigma_y = \frac{0.666e^y}{\sqrt{n}} \qquad (13\text{-}8)$$

Since for T = 10 the reduced variable in a type I extremal distribution is 2.25 (see Chapter 8) then the confidence limits are given by

$$2.25 \pm 6.33/\sqrt{n} \qquad (13\text{-}9)$$

Table 13-1. after Dalrymple (12), gives the upper and lower confidence limits with the corresponding return periods for various values of n.

Table 13-1

Confidence Limits for Index-Flood Homogeneity Test[1]

| Sample Size n | Lower Limit $y-2\sigma_y$ | $T_L$ | Upper Limit $y+2\sigma_y$ | $T_U$ |
|---|---|---|---|---|
| 5 | -0.59 | 1.2 | 5.09 | 160 |
| 10 | 0.25 | 1.8 | 4.25 | 70 |
| 20 | 0.83 | 2.8 | 3.67 | 40 |
| 50 | 1.35 | 4.4 | 3.15 | 24 |
| 100 | 1.62 | 5.6 | 2.88 | 18 |
| 200 | 1.80 | 6.5 | 2.70 | 15 |
| 500 | 1.97 | 7.7 | 2.53 | 13 |
| 1000 | 2.05 | 8.3 | 2.45 | 12 |

[1]From Dalrymple (12)

The procedure used for the test is to first of all plot $T_L$ and $T_U$ from Table 13-1 versus n on probability scale graph paper. Then, for each station in the region to be tested, the ratio of the 10-year event to the mean annual event is computed and an average ratio for the region calculated. Then the average ratio for the region is multiplied by the mean annual event for each station to give a modified 10-year event magnitude for each station. The return periods corresponding to these modified 10-year events are then found for each station from the individual station frequency curves, say $T_E$. The effective period of record of each gauging station is determined as the number of recorded annual events plus one half the number of events computed for that station by inter-station correlation, say $N_E$. Next, the coordinate pairs ($T_E$, $N_E$) for each station are plotted on the test graph showing curves of $T_L$ and $T_U$. Any station for which the plotted point is outside the confidence limit curves is then excluded from the homogeneous region. Figure 13-1 is a base graph which could be photo-copied for use in this test.

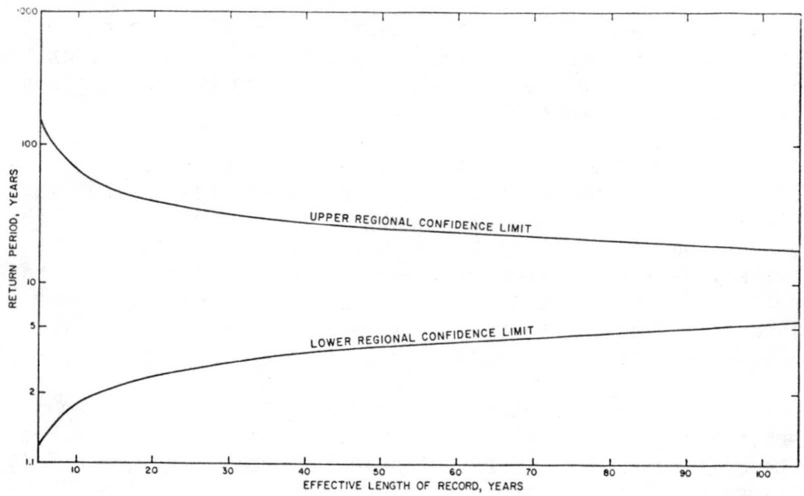


Figure 13-1. Plot to be used in regional homogeneity test.

For each station which remains in the hydrologically homogeneous region, ratios of events of different return periods to the mean annual event are computed for T values of say 1.1, 1.5, 5, 10, 20, 50 and median values of these ratios are determined for the region. A plot of these median ratios versus return period is then the regional frequency curve and represents the most likely relationship for all parts of the region.

The next major step in the index-flood analysis is to plot drainage area versus mean annual event for those stations within the homogeneous region and graphically fit a smooth curve through the points. An alternative available at this stage is to develop a multiple regression equation between mean annual event and basin characteristics.

To define a frequency curve at any location within the homogeneous region the mean annual event is determined from the curve relating this event to drainage area. The mean annual event is then multiplied by the median ratios for the various return periods required, as determined from the regional frequency curve.

Regional index-flood studies have been carried out for most of the states in the United States and for the South Saskatchewan River Basin (6), New Brunswick - Gaspé (7) and Nova Scotia (9) areas in Canada.

Benson (3) has noted three deficiencies found in the index-flood method:

(a) The index-flood (mean annual flood) for stations with short periods of record may not be typical. This means that the ratios of flood of different return periods to the index-flood may vary widely between stations.

(b) The homogeneity test is used to determine whether the differences in slopes of frequency curves are greater than may be attributed to chance alone. This test uses the ratio of the 10-year flood because many individual records are too short to adequately define the frequency curve at higher levels. It has been found in some studies that although homogeneity is apparently established at the 10-year level, the individual curves show wide and sometimes systematic differences at higher levels.

(c) In the use of the index-flood method, it has been accepted that within a flood-frequency region, frequency curves may be combined for all sizes of drainage areas, excluding only the largest. Although the variation in the slope of the frequency curve with drainage area had been investigated at the time of each study, it was studied at the 10-year point where the effect is small. The error of neglecting this drainage-area effect has been reduced by giving separate and special treatment to large streams. Recent studies by the USGS, for which ratios of less frequent floods were used, have shown, in all regions where such data are available, that the ratios of any specified flood to the mean annual flood will vary inversely with the drainage area. In general, the larger the drainage area, the flatter the frequency curve. The effect of drainage area is relatively greater for floods of higher recurrence intervals.

In applying the index-flood technique in Canada, Collier (6) has provided comments on some of the problems he encountered. It has been found in the foothills area of Alberta that normally the annual flood is due to snowmelt in the high regions usually combined with rainfall from Pacific air masses moving over the mountains from the west. Occasionally, however, moist tropical air is sucked up from the southeast and heavy precipitation occurs on the foothills. Simple flood frequency analysis does not distinguish between these two types of events and, as a result, a plot of the floods on arithmetic-normal probability paper shows a distinct S-shape.

Other comments made by Collier (6) included a discussion of the steps to take when a group of stations fails the regional homogeneity test as well as a discussion of the validity of the index-flood approach.

## Multiple Regression Techniques

*Estimating Events at Specific Return Periods* - Rather than plotting drainage area versus mean annual flood, as in the index-flood method, many investigators (see Benson (3) for a partial list) have studied the relationships between discharges at specified return periods and basin characteristics. In general, the relationships take the form

$$Q_T = f(A^a, B^b, C^c, \ldots, Z^z) \qquad (13\text{-}10)$$

where A, B, C,...,Z are independent variables and a, b, c,..., z are constants derived by multiple regression analysis. Since a similar relationship, but with different exponents, will be found for the mean annual flood, the ratio of $Q_T$ to the mean annual flood will not be a dimensionless constant as assumed in the index-flood method, but will be proportionate to the basin characteristics. A further advantage of this multiple regression modification to the index-flood method is that it obviates the necessity of assuming any underlying distribution for the flood peaks.

Many different procedures are available for determining the relevant parameters in Equation 13-10 including the simple linear regression, multiple linear regression, forward, backward and stepwise procedures. Packages of computer programs for these procedures are commonly available as library functions at computer centres. Wampler (31) has tested and compared more than 20 linear least squares computer programs.

De Coursey (13) used a canonical correlation procedure to select watershed characteristics and then derived a multiple regression matrix relationship

$$Q = aA + b \qquad (13\text{-}11)$$

where Q is a column vector of peak flows at various return periods, A is a column vector of watershed characteristics and a and b are respectively a matrix and a vector of regression coefficients. This approach preserves the intercorrelation between the dependent variables in vector Q.

*Estimating Regional Distribution Parameters* - An alternate approach is to regress the basin characteristics not with flood magnitudes at given return periods, but with the parameters of a chosen probability distribution. The relevant parameters are first derived for each of the individual sites in the region using single-station frequency analysis techniques and the regression equations are developed to estimate the distribution parameters at the required location. As an example, if the log-normal probability distribution is used (16) then the mean and standard deviation of the annual events should be used in the correlation.

Thomas and Benson (28) examined both techniques of estimating specific return periods and regional distribution parameters using over 70 parameters of streamflow including low flow peaks and flood volumes, parameters of annual and monthly means and flow distribution statistics.

These streamflow characteristics were each regressed against up to 31 basin characteristics including such parameters as drainage area, slopes, storage indices, precipitation totals and intensity, evaporation and temperature. The study showed that the most important parameters were drainage areas and mean annual precipitation. As might be expected the equations for characteristics of the mid-range of flows were more accurate for low flows. No significant relationships were found for coefficients of skew or serial correlation coefficients.

Square-Grid Method

The definition of areal runoff in geographical areas exhibiting basically similar hydrological characteristics can be facilitated by employing the square-grid technique as proposed by Solomon et al. (26). As implied in the term square-grid this technique entails dividing the study area into a uniform grid, the squares of which are identified in cartesian coordinates. Each square-grid has associated with it a set of parameters such as the elevations of each corner and centre point, the percentage areas of the square covered by lakes, swamps, forests, urban areas, etc., soil types, indicator of bedrock, etc., derived from topographic maps and other information sources such as soil surveys. From this basic data other physiographic characteristics such as mean slope, azimuth of slope, barrier height in different directions, distance to the sea in different directions can be easily derived for each square. Those square-grids containing meteorologic or hydrometric stations will also have associated with them data on mean annual precipitation and temperature and streamflow respectively. Each grid-square also carries identification of up to four streamflow directions into and out of adjacent squares so that drainage basin data can be accumulated and averaged automatically.

The grid interval determines to a large extent the accuracy of the results since the finer the grid the more basic data are available. Nevertheless, for a given set of conditions, the gain in accuracy obtained by further decreasing the grid interval diminishes greatly beyond a certain value of the interval, and a further increase of the number of squares is not warranted. In general, the optimum grid interval is determined by the size of the area, the size of the individual drainage basins considered, the details of the available data, the computer characteristics, the purpose and budget of the study, etc.. For usual problems, grid intervals between about 1 and 10 km can be considered. In Canada grid sizes of 10 km, 5 km and 4 km have been used in different studies. For the mountainous areas of British Columbia, physiographic data have been abstracted using a 2 km grid interval.

The steps involved in estimating the distribution of mean annual runoff using the square-grid method are as follows:

(a) Using data at selected meteorological stations a regression equation is established between mean annual temperature at the stations and corresponding square-grid physiographic characteristics. This equation is used with the data file of square-grid physiographic characteristics to estimate the mean annual temperature in the remaining squares. A similar analysis is used to make a preliminary estimate of mean annual precipitation in each square.

(b) Preliminary evaporation is estimated for each square using temperature, preliminary precipitation and Turc's equation (29)

$$E = P/(0.9 + P^2/L^2)^{\frac{1}{2}} \qquad (13\text{-}12)$$

where E is evaporation in mm, P is precipitation in mm, T is temperature in degrees Centigrade, and L is defined as

$$L = 300 + 25T + 0.5T^2 \qquad (13\text{-}13)$$

Using this equation, E = P for $P^2/L^2$ less than 0.1 $mm^2$.

(c) The preliminary estimates of precipitation and evaporation as described above are used to calculate a preliminary value of mean annual runoff for each square of the watershed having flow data as:

$$\text{Runoff} = \text{Precipitation} - \text{Evaporation} \qquad (13\text{-}14)$$

(d) The square-grid runoff as established under step (c) is used to compute preliminary average runoff for each basin having flow records, and, coefficients (K) representing the ratio between the recorded and computed average flow are established.

(e) A new precipitation value is computed for each square of the basins having flow data, using K as a correction factor:

$$\text{Precipitation (corrected)} = K \times \text{Runoff} + \text{Evaporation} \quad (13\text{-}15)$$

The entire error is attributed to precipitation.

(f) Using the corrected values of the precipitation in each square and the precipitation data at rain gauging stations, a new regression equation is established between precipitation and physiographic factors. Data at rain gauging stations is weighted 10 times larger than the precipitation estimates in ungauged squares.

(g) The procedure is then repeated as often as is required to obtain K values as close to 1 as is considered reasonable.

(h) Once square-grid runoff, precipitation and evaporation values are finalized in gauged basin areas, the runoff and precipitation distributions in ungauged areas are estimated using regression equations between the final precipitation and runoff in gauged areas and physiographic data.

The end result of the square-grid technique is a data file of mean annual temperature, precipitation and runoff for each grid square. Thus estimates of these parameters have been transferred from gauged to ungauged basins using physiographic characteristics as the transfer media. Solomon et al. (26) originally applied the square-grid method to Newfoundland but it has since been extended to cover all of Canada except for northern Ontario and the Arctic Archipelago (27).

Pentland and Cuthbert (25) have described a method by which the square-grid technique was extended for the generation of synthetic streamflow traces. The operational hydrology model proposed by Young and Pisano (35) was used because of the small number of statistical parameters required. This model uses only estimates of a single variance-covariance matrix and a single lag 1 covariance matrix whereas most other models require monthly matrices.

Young and Pisano's model (35) is set up as follows:

(a) The available streamflow data are logarithmically transformed.

(b) The transformed data are standardized on a monthly basis to eliminate the annual periodicity.

(c) The generating equation

$$X_{i+1} = AX_i + Be \quad (13\text{-}16)$$

is then used.

For m stations, $X_{i+1}$ and $X_i$ are standardized flows in successive time periods (m x 1 matrices). A and B are m x m matrices to be defined and e is an m x 1 matrix of random components.

(d) The matrix $M_o$ is the variance-covariance matrix, and $M_1$ is the covariance matrix with a lag of 1 time period.

(e) The matrix A can be defined from the equation

$$M_1 = AM_o \tag{13-17}$$

(f) The matrix B can be calculated by solving the equation:

$$B\,B^T = M_o - M_1\,M_o^{-1}\,M_1^T \tag{13-18}$$

(g) After having generated standardized variables, the data are destandardized, and the inverse logarithmic transform applied.

If one or more of the stations has no recorded data, monthly means and standard deviations can be estimated by regression analysis with basin physiographic characteristics derived from the square-grid data.

$$\mu = K_1 A^{a_1}\, B^{b_1}\, C^{c_1} \ldots\, Z^{z_1} \tag{13-19}$$

$$\sigma = K_2 A^{a_2}\, B^{b_2}\, C^{C_2} \ldots\, Z^{z_2} \tag{13-20}$$

where $\mu$ and $\sigma$ are the monthly mean streamflow and standard deviation, A, B,..., Z are physiographic characteristics and $K_1$, $K_2$, $a_1$, $b_1$,...,$z_1$ $a_2$,$b_2$,..., $z_2$ are regression constants. In order to estimate covariances for the streams with no recorded data, a multiple regression equation is established between covariances (representing cross correlations between stations) and the differences in physiographic characteristics for all pairs of gauged streams in the region.

In the covariance matrix with a lag of one time period, elements of the diagonal (representing serial correlation for each stream) for gauged stations can be calculated directly, and can be estimated for ungauged stations by regression analysis. The remainder of this matrix can be estimated as the product of its diagonal and the variance covariance matrix calculated earlier.

Pentland and Cuthbert (25) tested the generation procedure on five streams in the northeast of the Province of New Brunswick. Comparisons of simulated and recorded monthly means, standard deviations, serial and corss correlations and firm flows showed good agreements.

A procedure for using the square-grid technique to estimate events at required return periods on ungauged streams has been described by Kite (16) for the Mackenzie River area, Northwest Territories. Basically, equations similar to 13-19 and 13-20 were developed relating the mean and standard deviation of the annual maximum instantaneous flows to the square-grid physiographic data for those streams which are gauged. The relationships developed were then extended to ungauged streams and, assuming a lognormal probability distribution of annual extremes, event magnitudes at any required return period were calculated. Any other probability distribution thought suitable could have been used in place of the lognormal.

Kouwen (18) has described an advanced model, based upon square-grid techniques, used for the simulation of complete watersheds. The model allows hydrographs based on weather forecasts to be incorporated, for prediction of flood peaks. The basic input to the model consists of topographic data such as streambed elevations and landslope, drainage channel directions, watershed boundary coordinates, precipitation records, streamflow records and a soil permeability index.

The program is also set up in such a way that precipitation data from radar, and snow pack and soil moisture measurements from satellite can be included. The principal characteristic of the simulation is that runoff passes through successive 1 km x 1 km square elements from higher to lower elevations. For each element there exist relationships between channel capacity and drainage area, surface storage and channel inflow, surface storage and infiltration, subsurface storage and channel inflow, and channel storage and channel discharge.

### Use of Standard Frequency Distributions

In order to remove the necessity for personal judgement in drawing the preliminary frequency curves and to provide a means of computing confidence limits on the regional frequency curve in the index-flood method, Collier (6) produced an alternative regional analysis procedure. This procedure is recommended for regions where the Gumbel (type I extremal) distribution produces reasonably reliable individual flood frequency curves. The procedure is described briefly below:

(a) All stations in the region with 10 or more years of record (either natural flow or with minor regulation only) are selected. Stations with less than 10 years of record would usually be discarded, and most of the selected stations should have at least 15 years of record.

(b) A frequency curve covering the range up to the 100-year flood is constructed by the Gumbel (type I extremal) method for each of the individual stations. For the purpose of this discussion these will be called the preliminary curves. Confidence limits are constructed on each of the preliminary curves using the degree of confidence required in the regional curve.

(c) A homogeneity test is carried out exactly as in the index-flood method. For the purpose of this discussion it will be considered that each station passes the test and the region has therefore been demonstrated to be homogeneous.

(d) The preliminary curves are considered to be a sample composed of a number of different estimates of the same regional curve. The estimates are averaged by the method described in the next paragraph to obtain the required estimate of the regional curve.

(e) The averaging procedure can be carried out only if the preliminary curves are reduced to dimensionless terms (to remove the effect of the different sizes of the drainage basins concerned). This is accomplished by computing a set of flood ratios (ratio of flood to mean annual flood) for each of the stations over a range of arbitrarily selected recurrence intervals. The data for computing the ratios are read from the preliminary curves.

(f) For each of the selected recurrence intervals, the mean of the ratios from all the stations is computed. The resulting means are the flood ratios for the regional curve. These are plotted on Gumbel paper (with arithmetic ordinate scale) and the best-fit straight line drawn through them. The resulting line is taken as the required regional frequency curve.

(g) To compute confidence limits for the regional curve, a recurrence interval (say 50 years) is selected arbitrarily and the width of the confidence band at this interval is read off each of the preliminary curves. The width is taken as the vertical distance between the preliminary curve and the upper (or lower) band and it is expressed in cfs. The resulting figures are divided by the appropriate mean annual floods to produce a set of ratios, which are defined as the "errors" in the individual curves. The errors are combined by computing the square root by the number of stations. The resulting ratio is taken as the "error" in the regional curve or, in other words it is the width of the confidence bands for the regional curve at the selected recurrency interval (50 years in this case). The procedure is repeated at another recurrence interval (say 5 years). The "errors" from the two sets of computations are plotted on the regional curve by laying them off at the appropriate recurrence intervals in a vertical direction either side of the main curve. The resulting points are joined by straight lines to produce the required confidence bands for the regional curve. Note that these bands represent the same degree of confidence as was used in computing the confidence limits for the preliminary curves.

(h) Having obtained the dimensionless regional frequency curve, complete with confidence limits, it is necessary to introduce a relationship between mean annual flood and basin characteristics so that estimates may be made for ungauged drainage basins. In a study of the Province of Nova Scotia using Collier's procedure, Coulson (9) ran stepwise linear regressions of mean annual flood versus drainage area, size and position of lakes and swamps, main channel slope, average basin elevation, mean barrier elevation and mean annual precipitation. He ended up with an equation of the form

$$\bar{Q} = f(A_u + \lambda^k A_c) \tag{13-21}$$

where $A_u$ and $A_c$ are the drainage areas uncontrolled and controlled by lakes and swamps respectively,

$$\lambda = \frac{A_c - A_L}{A_c} \tag{13-22}$$

where $A_L$ is the total surface area of major lakes and swamps, and k is a constant optimized by minimizing the standard error of $\bar{Q}$, the mean annual flood.

Collier and Nix (7) used a similar approach in a flood frequency study of the New Brunswick-Gaspé region.

The principal advantage of Collier's alternative procedure for the regional frequency curve is that since no personal judgement is involved, the entire procedure can be programmed for computer. Although Collier (6) described the alternative procedure utilizing a type I extremal distribution there is no reason why any other type of distribution thought suitable could not be used.

Cruff and Rantz (11) have described the adaptations made by United States agencies to use the lognormal, extremal type I (Gumbel) and Pearson type III distributions in regional analysis. Basically the procedures used consist of the following steps:

(a) The mean and standard deviation of the peak discharge data at each gauging station are computed for the available periods of record. In the case of the lognormal distribution the means and standard deviations of the logarithms of the peak discharge data are computed in the procedure described by Cruff and Rantz (11) but this is not strictly necessary (see Chapter 6).

(b) The computed statistical parameters are then adjusted to a standard base period by computing linear correlations between concurrent peak discharges for a long-term station and the short-term stations. Then

$$\sigma_{1b} = \sigma_{1a} + (\sigma_{2b} - \sigma_{2a})\, R^2\, \sigma_{1a}/\sigma_{2a} \qquad (13\text{-}23)$$

and

$$\mu_{1b} = \mu_{1a} + (\mu_{2b} - \mu_{2a})\, R^2\, \sigma_{1b}/\sigma_{2b} \qquad (13\text{-}24)$$

where $\mu$ and $\sigma$ are the means and standard deviations respectively; subscript 1 refers to the short-term station, 2 to the long-term station, subscript a refers to the short-term period and b refers to the base period; and R is the coefficient of correlation between 1 and 2.

If these equations are derived by standardizing the means and standard deviations of the two periods of record at each station then the $R^2$ is not statistically correct and should be omitted.

In the case of the Pearson type III distribution, for which skews are needed, Equations 13-23 and 13-24 are used to generate events at the short-term stations to complete the record for all years of the base period by using the relationship:

$$x_1 = \mu_{1b} + R\,(x_2 - \mu_{2b})\, \sigma_{2b}/\sigma_{1b} \qquad (13\text{-}25)$$

where $x_1$ is the peak discharge to be estimated at a short-term station, $x_2$ is the peak discharge measured at the long-term station and the other parameters are as previously defined. When the full number of annual events is available for each of the short-term stations the coefficients of skew are computed for each station.

(c) The parameters of the distribution (mean, standard deviation and, for Pearson type III, skewness) are then related to the basin and climatologic characteristics by multiple linear regression equations as explained in section 13-3.

(d) For any site the mean, standard deviation (and, if necessary, coefficient of skew) can then be determined from the derived regression equations, and the event magnitude at return period T can be obtained from

$$x_T = \mu + K\,\sigma \qquad (13\text{-}26)$$

where K is the frequency factor. As explained in Chapter 3 the frequency factor can be developed in terms of T for each distribution and tables are commonly available.

## Regional Record Maxima

Suppose that there exists a set of n independent identically distributed concurrent series each containing k extreme events $x_{ij}$ i=1,n; j=1,k. If the maximum event of each of the n series is abstracted and ordered from highest to lowest in a new series, $y_i$, i=1,n then the probability, $P(y > y_i)$, that another event y exceeds the ith event in the series of maxima $y_i$ is given by Conover and Benson (8) as:

$$P(y > y_i) = \sum_{m=o}^{i-1} n!/[(n-m)!\ k \prod_{j=o}^{m} (n+1/k-j)] \qquad (13\text{-}27)$$

As an example, Garrigan (5) has shown that for the three series of four events, $x_{ij}$,

| | | |
|---|---|---|
| 3 | 69 | 3 |
| 38 | 24 | 48 |
| 17 | 61 | 60 |
| 32 | 30 | 83 |

the series of maximum events, $y_i$, (83, 69, 38) and the probability, $P(y > 69)$, that another event y exceeds the second largest event in the series of maxima is

$$P(y > 69) = \sum_{m=o}^{1} 3!/[(3-m)!\ 4 \prod_{j=o}^{m} (3+1/4-j)] \qquad (13\text{-}28)$$

$$P(y > 69) = 0.180 \qquad (13\text{-}29)$$

An analogy which can be made here is with annual maximum streamflows recorded at a set of gauging stations within a hydrologically homogeneous region. Within this homogeneous region it is a reasonable assumption that the same probability distribution is applicable to the records of maximum events on each stream. Just as in the index-flood method the different records could be reduced to an identical distribution by normalizing with the computed mean annual floods. The procedure outlined above then takes the n independent samples of k events and forms a sample size nk; thus probabilities can be computed associated with return periods of nk years instead of only n years. The catch is that streamflow records are not independent but are quite strongly cross-correlated. This reduces the maximum return period available from nk to f(R)nk where f(R) is some function of the correlation coefficient, R, between streamflow records. The expressions f(R) vary between 1 when R=0 (independent records) to 1/n when R=1 (identical records).

The probability of another random event, y, exceeding one of the ordered record maxima, $y_i$, cannot be determined analytically when the records are not independent. By assuming that the exceedence probability is independent of the identical distribution of the records, Carrigan (5) has derived the probability by data generation. Using the normal distribution for simplicity the generation model used by Carrigan is

$$X = B\varepsilon \qquad (13\text{-}30)$$

where X is an n x k matrix of generated events, B is an n x n principal component matrix and ε is an n x k matrix of independent normally distributed random numbers with zero mean and unit variance. The principal component matrix B is derived from the correlation matrix R of the n records of k hydrologic events as follows:

$$B = E\lambda \qquad (13\text{-}31)$$

where E is an n x n matrix of eigenvectors obtained from R, and λ is an n x n matrix for which the diagonal elements are the square roots of the eigenvalues of R and the off-diagonal elements are zero.

After generation of X, the n maxima are selected and put in order of magnitude and the exceedence probabilities computed using a digital approximation to the normal distribution.

In summary, the method extracts from the records of a series of gauging stations a matrix of the inter-station correlation. This correlation is then incorporated into a large number of generated events from which extreme probabilities can be measured. In effect the method converts the spatially-distributed information into time-distributed information on extreme events.

## Bayesian Analysis

A further method of combining a single station analysis with regional analysis information is the use of Bayesian analysis. Bayes theorem states that the posterior probability distribution function (pdf) is proportional to the product of the pdf containing the prior information (e.g. from regional analysis) and the likelihood function of the sample pdf. In order to provide a tractable posterior pdf it is necessary to choose a prior pdf and a likelihood function which will conveniently combine. Vicens et al. (30) used a normal distribution for the mean annual flows of a New Hampshire river and a normal inverted 2-parameter gamma distribution for the prior pdf. For extreme events a better distribution to use might be the lognormal. Given a sample of n annual maximum events $x_i$, i=1, n then the sufficient statistics of the normal distribution of $y = \log_{10} x$ are the mean $\bar{y}$, the variance $s^2$, the sample size n, and the number of degrees of freedom, $v = n-1$. As an example, Table 13-2, lists the sufficient statistics for the Tobique River at Plaster Rock, New Brunswick, for the period 1955-1966. Wood and Rodriguez-Iturbe (32) have shown other examples.

The same statistics can then be defined for the prior distribution as:

$$\bar{y} = E[\mu] \qquad (13\text{-}32)$$

$$S'^2 = [(v'-2)/v']\, E[\sigma^2] \qquad (13\text{-}33)$$

$$n' = E[\sigma^2]/V[\mu] \qquad (13\text{-}34)$$

$$v' = 2E^2[\sigma^2]/V[\sigma^2] \qquad (13\text{-}35)$$

where $E[\mu]$, $V[\mu]$, $E[\sigma^2]$, $V[\sigma^2]$ are the expected value and variance of the mean and variance of the prior distribution. These statistics may be obtained by regressions against physiographic or meteorologic variables. For example, using data from Collier and Nix (7), a regional analysis was carried out using 19 basins in Gaspé, New Brunswick and northern Maine with periods of record varying from 12-40 years. Independent variables used included drainage area, area of lakes and mean slope of basin. The resulting statistics are shown in Table 13-2. Note that n' and v' from the prior pdf can be interpreted as the number of years of data-equivalent for the mean and variance contained in the prior pdf (Vicens et al. (30)).

The regional information in the prior distribution and the sample information in the single station record are then combined in Bayes theorem to yield a posterior pdf. The statistics of this posterior pdf are given by

$$\bar{y}'' = (n'\bar{y}' + n\bar{y})/(n' + n) \qquad (13\text{-}36)$$

$$S''^2 = (v's'^2 + n'\bar{y}'^2 + vs^2 + n\bar{y}^2 - n''\bar{y}''^2)/v'' \qquad (13\text{-}37)$$

$$n'' = n' + n \qquad (13\text{-}38)$$

$$v'' = v' + v + 1 \qquad (13\text{-}39)$$

Table 13-2

Bayesian Analysis, Tobique River, New Brunswick

| Statistic | Sample Value | Prior Value | Posterior Value |
|---|---|---|---|
| n, years | 12 | 19 | 31 |
| v, years | 11 | 21 | 33 |
| $\bar{y}$, log cfs | 4.0674 | 4.1500 | 4.1180 |
| $s^2$, log (cfs)$^2$ | 0.0579 | 0.0312 | 0.0417 |

The statistics for the posterior pdf of the Tobique river are also given in Table 13-2. Figure 13-2 shows the frequency curves from the sample and from the posterior distribution.

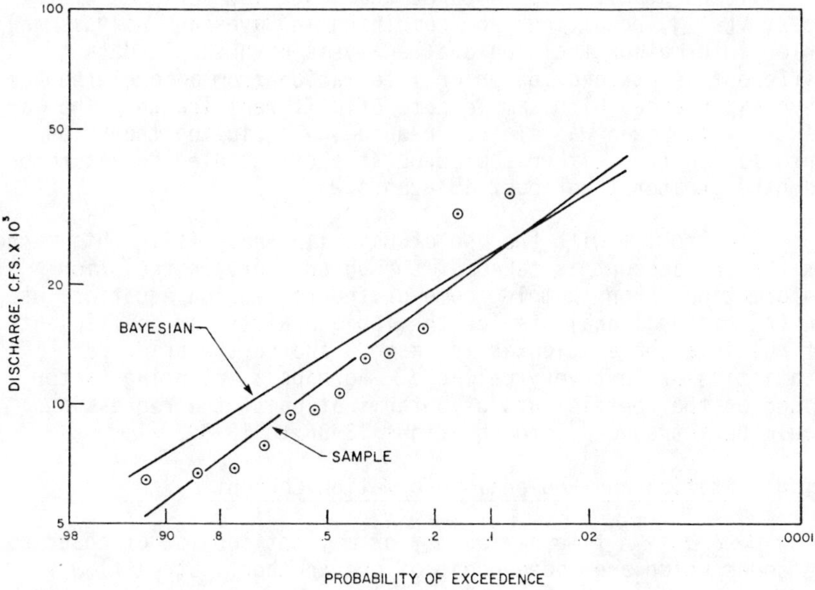


Figure 13-2. Comparison of sample and Bayesian frequency curves, Tobique River at Plaster Rock.

The prior information does not necessarily have to come from a regional analysis; Vicens et al. (30) also give an example of prior information obtained from a simple watershed model. Given the four sufficient parameters of the total annual precipitation distribution (assumed to be independent normal) and a simple model such as:

$$q_i = (1-b)\ x_i \tag{13-40}$$

where $q_i$ is the runoff in year i, b is the "loss" and $x_i$ is the annual precipitation in the same units as $q_i$, then the sufficient parameters of the runoff distribution are

$$E[\mu'] = (1 - E[b])\ E[\mu_x] \tag{13-41}$$

$$V[\mu'] = E^2[\mu_x]\ V[b] + (1 - E[b]^2)\ V[\mu_x] \tag{13-42}$$

$$E[\sigma'^2] = \{(1 - E[b])^2 + V[b]\}\ E[\sigma_x^2] \tag{13-43}$$

$$V[\sigma'^2] = 4(1 - E[b])^2\ E^2[\sigma_x^2]\ V[b] + (1 - E[b])^4\ V[\sigma_x^2] \tag{13-44}$$

assuming that all covariances are zero. The hydrologist must then estimate E[b] and V[b] before the Bayesian analysis can be used to combine the information from the model with the information from the sample to give the posterior information.

Vicens et al. have compared the posterior variance of the mean, $V[\mu'']$, resulting from combining in Bayesian analysis (a) prior information from regional analysis with sample data of different length and (b) prior information from a model (subjective experience) with sample data of different length. The variance of the mean was clearly reduced by including the prior information from either source but the effect died out at record lengths greater than about 40 events.

One problem with the use of Bayesian analysis in this manner is that no account is taken of the uncertainty in the prior information. For example, some of the regression equations used in the regional analysis for the Tobique River had coefficients of multiple correlation as low as 0.3 and so the prior pdf statistics are not very reliable. Perhaps a weighting factor based on the coefficients of determination of the regressions could be introduced into Equations 13-36 to 13-39.

## Single Station and Regional Information Content

*General* - A time series may or may not consist of observed outcomes which are independent of one another. Streamflow is a hydrologic variable whose observations, equally spaced in time, are not necessarily randomly distributed. Because of natural storage such as groundwater, lakes, swamps and annual persistance as well as manmade factors such as reservoirs, the stochastic precipitation variable becomes modulated and serially correlated. This means that each unit of streamflow data does not contain totally new and independent information; units tend to repeat some of the previously obtained information. A data set of N units may therefore only contain a lesser number, $N_e$, of effective data units.

The early stage of any regional analysis procedure calls for the examination of available data. It is usual that some gauging station records will be longer than others within the required region and often, after selection of a base period of record, it will be necessary to fill in gaps in some records and extend other records to the full base period. This provision of missing data can have two general purposes: (a) To provide estimates of event magnitudes in order to better obtain plotting positions of recorded events. This procedure is used in the index-flood regional analysis technique where the estimated event magnitudes are not themselves used at all, they are there merely to improve estimates for the recorded events. (b) To improve estimates of the parameters of a theoretical distribution of recorded events such as the mean and standard deviation.

Data are not only serially correlated, but because streamflows in rivers within a region are affected by the same conditions of precipitation and radiation, simultaneous observations of streamflow in different rivers will not be independent observations but will contain an overlap of information. Thus a region of n gauging stations may reduce to a much smaller number, $n_e$, of effective stations or equivalent independent stations.

*Single Station Information Content* - The purpose of maintaining records of precipitation, stage, streamflow, etc., is to extract from the recorded observations information on the parameters of the underlying distribution. Matalas and Langbein (21) defined the amount of information given by a statistical estimate, I, as the reciprocal of the variance of the estimate. Considering the mean, $\mu$, of a random series of N events, $x_i$, i=1...n,

$$\mu = \sum_{i=1}^{N} x_i/N \tag{13-45}$$

an estimate of the variance of $\mu$ is given by

$$\text{var}\ \mu = \sigma^2/N \tag{13-46}$$

where $\sigma^2$ is an estimate of the population variance of the random time series. Defining the random series as the standard, the relative information content about the mean of any other time series with variance of the estimated mean, $\sigma_\mu^2$, is

$$I_\mu = (\sigma^2/N)/\sigma_\mu^2 \tag{13-47}$$

referred to the random series. Since variances are always positive $I_\mu$ can vary from zero to plus infinity. If $I_\mu$ is less than unity the time series being tested conveys less information about the mean than a random series of the same length. If $I_\mu$ is greater than unity then the time series contains more information about the mean than an equal length random series.

Many time series exhibiting persistance, such as streamflow, can be described by a simple first order linear Markov model (34) such as:

$$x_{i+1} = R_1 x_i + \varepsilon_{i+1} \tag{13-48}$$

where $x_i$ and $x_{i+1}$ are the variable values at time i and i+1 respectively, $\varepsilon_{i+1}$ is a random component independent of x, and $R_1$ is the first order serial or autocorrelation coefficient where, in general, the kth order autocorrelation coefficient is defined (33) as:

$$R_k = \frac{\sum_{i=1}^{n-k} (x_i-\mu)(x_{i+k}-\mu)}{(N-k)\sigma^2} \tag{13-49}$$

where N is the number of observations and $\mu$ and $\sigma$ are the sample estimates of the mean and standard deviation of the time series.

For a first order linear Markov model the variance of the mean, $\sigma_\mu^2$, is given (21) by

$$\sigma_\mu^2 = \frac{\sigma^2}{N}\left[\frac{1+R_1}{1-R_1} - \frac{2}{N}\,\frac{R_1\ (1-R_1^N)}{(1-R_1)^2}\right] \tag{13-50}$$

From Equation 13-47 the relative information content on the mean is

$$I_\mu = \left[\frac{1+R_1}{1-R_1} - \frac{2}{N}\,\frac{R_1\ (1-R_1^N)}{(1-R_1)^2}\right]^{-1} \tag{13-51}$$

which is less than unity for $R_1 > 0$.

If a number of independent observations of a random time series, $N_e$, contain the same information content about the mean as the number N observations of the Markov model then

$$I_\mu = N_e/N \tag{13-52}$$

and

$$N_e = N\left[\frac{1+R_1}{1-R_1} - \frac{2}{N}\,\frac{R_1\ (1-R_1^N)}{(1-R_1^2)}\right]^{-1} \tag{13-53}$$

*Two-Station Transfer of Information* - Inter-station transfer of information is commonly used in regional analysis to fill in missing data or to extend short time series to a longer common base period. The method used in the index-flood method of regional analysis (12) is as follows:

A graph is drawn of the flow at one station versus the flow at the other station for each year of the common period of record. A straight line is fitted by eye through the coordinate points and this line is then used to extend the shorter period of record. This process is a simplification of the least squares fitting of a linear regression equation of the type

$$y = mx + c \tag{13-54}$$

where x and y are the annual maximum flows at the two stations, m is the slope and c is the intercept of the straight line. The missing event magnitudes are then estimated from this regression line and the total events, recorded and estimated are placed in order of decreasing magnitude. Plotting positions (see Chapter 2) are assigned to the recorded events on the basis of the total number of events and the estimated events are then discarded and used no further.

Many investigators (19, 24) have found that the logarithms of hydrologic events are better correlated than the recorded events and have used equations such as

$$\ln y = m_1 \ln x + c_1 \tag{13-55}$$

where x and y may be the recorded events or the deviations of the recorded events from some mean value.

The question arises as to whether the estimated events actually increase the information content of the shorter time-series, i.e. does the extended data provide better estimates of the population distribution parameters than the originally recorded series? Langbein (19) has shown that to improve the significance of the mean of a time series the effective period of record, $N_e$, of a combined recorded and estimated record must be greater than $N_1$, the number of years of recorded data, where

$$N_e = \frac{N_1 + N_2}{1 + \frac{N_2}{N_1 - 2}(1 - R^2)} \tag{13-56}$$

$N_2$ is the number of years of estimated data and $R^2$ is the coefficient of determination of the simple linear regression used to provide the estimated data.

If two random normally distributed time series x, of length $N_1 + N_2$, and y, of length $N_1$, are linearly related with a simple linear correlation coefficient R, and the time series x is used to extend time series y by $N_2$ data points, then the variance, $\sigma_\mu^2$, of the weighted mean of series, y, $\mu_y$, where

$$\mu_y = \frac{N_1 \mu_1 + N_2 \mu_2}{N_1+N_2} \tag{13-57}$$

is given (21) by

$$\sigma_\mu^2 = \frac{\sigma_y^2}{N_1}\left[1 - \frac{N_2}{N_1 + N_2}\left\{R^2 - \frac{(1 - R^2)}{(N_1 - 3)}\right\}\right] \tag{13-58}$$

where $\mu_1$ and $\mu_2$ are the sample means of time series y based on $N_1$ observations and $N_2$ regression estimates respectively and $\sigma_y^2$ is the variance of time series y based on the $N_1$ observations.

Referring back to Equation 13-47 the information content on the mean of the extended time series y is seen to be:

$$I_\mu = \left[1 - \frac{N_2}{N_1 + N_2}\left\{R^2 - \frac{(1 - R^2)}{(N_1 - 3)}\right\}\right]^{-1} \quad (13\text{-}59)$$

and the effective number of observations, $N_e$, is given by:

$$N_e = (N_1 + N_2)\left[1 - \frac{N_2}{N_1 + N_2}\left\{R^2 - \frac{(1 - R^2)}{(N_1 - 3)}\right\}\right]^{-1} \quad (13\text{-}60)$$

For the cross-correlation to provide additional information on the mean, $I_\mu > 1$ and from Equation 13-59

$$\frac{- N_2 R^2}{N_1 + N_2} + \frac{N_2}{(N_1 + N_2)(N_1 - 3)} - \frac{N_2 R^2}{(N_1 - 3)} < 0 \quad (13\text{-}61)$$

from which

$$R^2 > \frac{1}{N_1 - 2} \quad (13\text{-}62)$$

Similarly, Fiering (14) concluded that correlation should not be used to augment time series for estimation of the variance unless the computed information content on the variance, $I_{\sigma^2}$, is greater than unity, where

$$I_{\sigma^2} = \left\{1 + \frac{N_2}{2(N_1 + N_2 - 1)^2}\left[2A(N_1 - 1) + (N_2 + 2)(N_1 - 1)B + (N_1 + N_2 - 1)(N_1 - 1)C + (N_1 + 1)(2N_1 + N_2 - 2)\right]\right\}^{-1} \quad (13\text{-}63)$$

in which

$$A = (N_1 - 1)R^4 + (N_1 + 4)R^2 (1 - R^2) + \frac{N_1 + 1}{N_1 - 3}(1 - R^2)^2 \quad (13\text{-}64)$$

$$B = R^4 + \frac{6R^2(1 - R^2)}{N_1 - 3} + \frac{3(1 - R^2)}{(N_1 - 3)(N_1 - 5)} \quad (13\text{-}65)$$

and

$$C = \frac{2(N_1 - 4)(1 - R^2)}{N_1 - 3} \quad (13\text{-}66)$$

Fiering (14) concluded that, in general, the estimate of the population variance will be improved if $R > 0.85$.

If the two time series x and y are not random but are serially correlated then for an equal number of observations, N, the effective number of data points, $N_e$, has been given by Yevjevich (34) as:

$$N_e = N/(1 + 2R_1R_1' + 2R_2R_2' + 2R_{n-1}R_{n-1}') \qquad (13\text{-}67)$$

where $R_k$ and $R_k'$ are the kth order autocorrelation coefficients of the x and y time series respectively. This equation is only useful if all periodicities have been removed from both time series. If the equal-length time series x and y can be described by firșt order linear Mrakov models then the effective number of observations is given by:

$$N_e = N\left[\frac{1 - R_1R_1'}{1 + R_1R_1'}\right] \qquad (13\text{-}68)$$

If x and y are two first order linear Markov models of length $N_1+N_2$ and $N_1$ respectively and x is correlated with y to provide $N_2$ regression estimates for y then the relative information content for the mean of the augmented time series varies with $N_1$, $N_2$, $R_1$, $R_1'$ and R in a complex fashion. Assuming that $R_1=R_1'$, Matalas and Langbein (21) have tabulated values of $I_\mu$ for different values of these variables.

Equations such as 13-54 and 13-55 yield event magnitudes on the regression line and, although this does not affect the estimate of the mean, it does induce a bias in the estimate of the variance. To overcome this bias it is necessary to introduce into the generating equation a random variable with mean zero and variance $(1 - R^2)\sigma^2$ where $\sigma^2$ is the variance of the recorded series. The true regression equation thus becomes:

$$y = \mu_y + m(x - \mu_x) + (1 - R^2)^{\frac{1}{2}} \sigma_y\varepsilon \qquad (13\text{-}69)$$

where $\varepsilon$ is a normally distributed (0,1) random number. The term $(1 - R^2)^{\frac{1}{2}} \sigma_y\varepsilon$ is the random component of the generated time series and in communications theory is referred to as noise.

Under the assumptions that (a) events are independently distributed in time, (b) the concurrent events for the two sequences have a joint normal distribution, (c) the relation between the concurrent events is defined by a linear regression, and (d) no changes occur in the hydrologic regimes with which the sequences are associated, Matalas and Jacobs (22) have evaluated the reliability of estimates of population parameters under conditions of noise and no-noise.

Matalas and Jacobs (22) recommended the use of the following equations to compute the mean, $\mu_y$, and variance, $\sigma_y^2$, of an extended time series:

$$\mu_y = \bar{y}_1 + \frac{N_2}{N_1 + N_2} m(\bar{x}_2 - \bar{x}_1) \qquad (13\text{-}70)$$

and

$$\sigma_y^2 = \frac{1}{N_1 + N_2 + 1} \Bigg\{ (N_1 - 1) S_{y_1}^2 + (N_2 - 1) m^2 S_{x_2}^2$$

$$+ (N_2 - 1) \alpha^2 (1 - R^2) S_{y_1}^2 + \frac{N_1 N_2}{(N_1 + N_2} m^2 (\bar{x}_2 - \bar{x}_1)^2 \Bigg\} \qquad (13\text{-}71)$$

where $\bar{y}_1$ and $S_{y_1}^2$ are the mean and variance of the recorded events in the augmented series, $\bar{x}_1$ is the means of the concurrent augmenting series, $\bar{x}_2$ and $S_{x_2}^2$ are the mean and variance of the total number of events in the augmeting series, and

$$\alpha^2 = \frac{N_2(N_1 - 4)(N_1 - 1)}{(N_2 - 1)(N_1 - 3)(N_1 - 2)} \qquad (13\text{-}72)$$

Equations 13-70 and 13-71 should only be used if the inter-station correlation coefficient, R, is greater than the critical values given in Tables 13-3, 13-4 and 13-5.

Table 13-3

Critical Minimum Values of R for Estimation of the Mean[1]

| N | 10 | 15 | 20 | 25 | 30 |
|---|---|---|---|---|---|
| R | 0.35 | 0.28 | 0.24 | 0.21 | 0.19 |

[1]From Matalas and Jacobs (22)

*Regional Transfer of Information* - As well as considering the transfer of information from a long-term station to an adjacent short-term station, hydrologists are often interested in studying how an hydrologic variable, such as river discharge, varies with the physical parameters describing the drainage area. The variation may be studied by assembling the data for many gauging stations and using regression analysis to define a relationship between the hydrologic variable and the physiographic characteristics.

In a given region, however, rivers may rise in response to a rain storm that affects all the rivers in the region and, at another time, may be low due to a common lack of rainfall. Thus the flows of different streams are affected by common causes and are therefore not independent but cross-correlated.

Table 13-4

Critical Minimum Values of R for Estimation of the Variance Including Noise Component[1]

| $N_2$ | $N_1$ = 10 | 15 | 20 | 25 | 30 |
|---|---|---|---|---|---|
| 10 | 0.65 | 0.54 | 0.52 | 0.42 | 0.38 |
| 15 | 0.65 | 0.54 | 0.51 | 0.42 | 0.39 |
| 20 | 0.65 | 0.54 | 0.51 | 0.42 | 0.39 |
| 25 | 0.65 | 0.54 | 0.50 | 0.42 | 0.39 |
| 30 | 0.65 | 0.54 | 0.50 | 0.42 | 0.39 |

Table 13-5

Critical Minimum Values of R for Estimation of the Variance Excluding Noise Component[1]

| $N_2$ | $N_1$ = 10 | 15 | 20 | 25 | 30 |
|---|---|---|---|---|---|
| 10 | 0.73 | 0.63 | 0.70 | 0.76 | 0.76 |
| 15 | 0.75 | 0.77 | 0.79 | 0.80 | 0.80 |
| 20 | 0.76 | 0.79 | 0.81 | 0.81 | 0.82 |
| 25 | 0.78 | 0.80 | 0.84 | 0.83 | 0.81 |
| 30 | 0.77 | 0.80 | 0.82 | 0.83 | 0.84 |

[1]From Matalas and Jacobs (22)

where $R_{ij}$ is the cross-correlation coefficient between stations and j.

If a number n of hydrometric gauging station records within a hydrologically homogeneous region are intercorrelated, then the effective number of stations or equivalent number of independent gauging stations, $n_e$ can be derived as follows (34): If the mean and variance of the observations at the jth gauging station are given by:

$$\mu_j = \frac{1}{N} \sum_{i=1}^{N} x_{ij} \quad (13\text{-}72)$$

$$\sigma_j^2 = \frac{1}{N-1} \sum_{i=1}^{N} (x_{ij} - \mu_j)^2 \quad (13\text{-}73)$$

where N is the number of observations at each station, then the estimates of the regional mean, $\mu$, and the variance of the regional mean, $\sigma_\mu^2$, are (33)

$$\mu = \frac{1}{N} \sum_{j=1}^{n} \mu_j \quad (13\text{-}74)$$

$$\sigma_\mu^2 = \frac{1}{n} \sum_{j=1}^{n} \sigma_j^2 + \frac{2}{n^2} \sum_{j=1}^{n-1} \sum_{i=j+1}^{n} R_{ij}\sigma_i\sigma_j \quad (13\text{-}75)$$

Equation 13-75 can be simplified by defining the regional mean cross-correlation coefficient, $\bar{R}$, as:

$$\bar{R} = \frac{2 \sum_{j=1}^{n-1} \sum_{i=j+1}^{n} R_{ij}}{n(n-1)} \quad (13\text{-}76)$$

If the time-series are standardized to a common mean of zero and variance $\sigma^2$ then Equation 13-75, incorporating Equation 13-76 becomes:

$$\sigma_m^2 = \frac{\sigma^2}{n} [1 + \bar{R}(n - 1)] \quad (13\text{-}77)$$

The relative information content on the regional mean is therefore given by:

$$I_{\mu} = [1 + \bar{R}(n - 1)]^{-1} \qquad (13\text{-}78)$$

and the effective number of stations, $n_e$, or equivalent number of independent stations is:

$$n_e = n/[1 + \bar{R}(n - 1)] \qquad (13\text{-}79)$$

In a study of regional flood frequency relations for 164 basins in New England, Benson (4) found that using Equation 13-79 the effective number of gauging stations or equivalent number of independent stations was 3.8. In a later study, Matalas and Benson (20) point out that, assuming the same value of $\bar{R}$, if n were only 20 stations, $n_e$ would be 3.4 and if n were 500, $n_e$ would be 3.8. This illustrates the rapid arrival at the limiting number of independent records. No appreciable increase in information is attained by using 500 stations instead of 20, if they are within the same region. As a further example, if there are an infinite number of stations and $\bar{R} = 0.1$, the effective number of stations is only 10.

The theory of regression analysis is based on the assumption, amongst others, that the values of the dependent variable are mutually independent. It is apparent, then, that in equations such as:

$$x = a_0 + a_1A + a_2B + a_3C + \ldots.. \qquad (13\text{-}80)$$

where x is some streamflow characteristic and A, B, C, ... are basin physiographic characteristics, the regression theory is impaired since the x values are not independent. Matalas and Benson (20) have investigated this problem very thoroughly and have concluded that the estimation of the regression constants, $a_0$, $a_1$, ... and the subsequent estimated value $\hat{x}$ are not affected by inter-station correlation. If inter-station correlation is present, however, the variance of $a_0$ will be larger than if there were no correlation, the variances of $a_1, a_2, \ldots, a_n$ will be smaller and the variance of $\hat{x}$ may be larger or smaller.

Considering a set of n gauging stations each of N observations but which are both serially and cross-correlated then Equations 13-67 and 13-79 can be combined to give a total effective number of station-events defined as:

$$N_e n_e = Nn/\left\{\left[1 + \bar{R}(n - 1)\right] (1 + 2\bar{R}_1 + 2\bar{R}_2 + \ldots 2\bar{R}_n)\right\} \qquad (13\text{-}81)$$

where $\bar{R}_1$, $\bar{R}_2$ are the average serial correlation coefficients of the n time series.

Developing this analysis, the procedure can be used as a means of defining homogeneous hydrologic regions (33, 17).

References


1. Alexander, G. N., 1954, Some Aspects of Time Series in Hydrology, J. Inst. Engineers (Australia) p. 196.

2. Benson, M. A., 1960, Characteristics of Frequency Curves Based on a Theoretical 1,000-Year Record, USGS Water Supply Paper 1543-A, pp. 51073.

3. Benson, M. A., 1962, Evolution of Methods for Evaluating the Occurrence of Floods, USGS Water Supply Paper 1580-A.

4. Benson, M. A., 1962, Factors Influencing the Occurrence of Floods in a Humid Region of Diverse Terrain, USGS Water Supply Paper 1580-B.

5. Carrigan, P. H., Jr., 1971, A Flood-Frequency Relation Based On Regional Record Maxima, USGS Professional Paper No. 434-F.

6. Collier, E. P., 1963, Regional Flood Frequency Analysis, unpublished paper, Water Resources Branch, Ottawa.

7. Collier, E. P. and G. A. Nix, 1967, Flood Frequency Analysis for the New Brunswick-Gaspé Region, Technical Bulletin No. 9, Inland Waters Branch, Ottawa.

8. Conover, W. J. and M. A. Benson, 1963, Long-Term Flood Frequencies Based on Extremes of Short-Term Records, USGS Professional Paper No. 450-E, pp. E159-E160.

9. Coulson, A., 1967, Flood Frequencies of Nova Scotia Streams, Technical Bulletin No. 4, Water Resources Branch, Ottawa.

10. Coulson, A., 1967, Estimating Runoff in Southern Ontario, Technical Bulletin No. 7, Inland Waters Branch, Ottawa.

11. Cruff, R. W. and S. E. Rantz, 1965, A Comparison of Methods Used in Flood Frequency Studies for Coastal Basins in California, USGS Water Supply Paper 1580-E.

12. Dalrymple, T., 1960, Flood Frequency Analyses, USGS Water Supply Paper 1543-A.

13. De Coursey, D. G., 1973, Objective Regionalization of Peak Flow Rates, Proceedings of Second International Symposium in Hydrology, pp. 395-405, Water Resources Publications, Fort Collins, Colorado.

14. Fiering, M. B., 1963, Use of Correlation to Improve Estimates of the Mean and Variance, USGS Professional Paper No. 434-C.

15. Kalinin, G. P., 1960, Calculation and Forecasts of Streamflow from Scanty Hydrometric Readings, Trans. Interregional Seminar on Hydrologic Networks and Methods, Bangkok, 1959, WMO Flood Control Series No. 15, pp. 42-52.

16. Kite, G. W., 1974, Case Study of Regional Analysis Techniques for Design Flood Estimation, Can. J. Earth Sciences, Vol. 11, No. 6, pp. 801-808.

17. Kite, G. W., 1973, Serial Correlation as a Measure of Regional Uniformity, unpublished notes, Water Resources Branch, Ottawa.

18. Kouwen, N., 1973, Watershed Modelling Using a Square-Grid Technique, Proceedings of the 9th Canadian Hydrology Symposium, Edmonton, Alberta.

19. Langbein, W. B., 1960, Hydrologic Data Networks and Methods of Extrapolating or Extending Available Hydrologic Data, Trans. Interregional Seminar on Hydrologic Networks and Methods, Bangkok, 1959, WMO Flood Control Series No. 15, pp. 13-41.

20. Matalas, N. C. and M. A. Benson, 1961, Effect of Interstation Correlation on Regression Analysis, Journal of Geophysical Research, Vol. 66, No. 10, pp. 3285-3293.

21. Matalas, N. C. and W. B. Langbein, 1962, Information Content of the Mean, Journal of Geophysical Research, Vol. 67, No. 9, pp. 344-348.

22. Matalas, N. C. and B. Jacobs, 1964, A Correlation Procedure for Augmenting Hydrologic Data, USGS Professional Paper No. 434-E.

23. Matalas, N. C., 1967, Time Series Analysis, Water Resources Research, Vol. 3, No. 3, pp. 817-830.

24. Pentland, R. L., 1967, Extending Streamflow Records, Program No. 20, unpublished paper, Water Resources Branch, Ottawa.

25. Pentland, R. L. and D. R. Cuthbert, 1971, Operational Hydrology for Ungauged Streams by the Grid Square Technique, Water Resources Research, Vol. 7, No. 2, pp. 283-291.

26. Solomon, S. I., T. P. Denouvilliez, C. Cadou and E. J. Chart, 1968, The Use of a Square-Grid System for Computer Estimation of Precipitation, Temperature and Runoff in a Sparsely Gauged Area, Water Resources Research, Vol. 4, No. 5, pp. 919-930.

27. Solomon, S. I. and A. S. Qureshi, 1972, Hydrologic Data Banks - Present Status and Potential, Engineering Journal, Vol. 55, No. 1/2, pp. 9-14.

28. Thomas, D. M. and M. A. Benson, 1969, Generalization of Streamflow Characteristics from Drainage Basin Characteristics, USGS Open-file Report, Washington, D. C., 45 p.

29. Turc, L., 1954, Le Bilan D-eau des Sols: Relations entre les Percipitations, L'évaporation et L'écoulement, Annales Agronomiques, Vol. 4, pp. 491-595.

30. Vicens, G. J., Rodriguez-Iturbe, I. and J. C. Schaake, Jr., 1975, A Bayesian Framework for the Use of Regional Information in Hydrology, Water Resources Research, Vol. 11, No. 3, pp. 405-414.

31. Wampler, R. H., 1969, An Evaluation of Linear Least Squares Computer Programs, Journal of Research of the National Bureau of Standards - B. Mathematical Sciences, Vol. 73B, No. 2, pp. 59-90.

32. Wood, E. F. and I. Rodriguez-Iturbe, 1975, Bayesian Inference and Decision Making for Extreme Hydrologic Events, Water Resources Research, Vol. 11, No. 4, pp. 533-542.

33. Yevjevich, V. M., 1964, Fluctuations of Wet and Dry Years, Part II, Analysis by Serial Correlation, Hydrology Paper No. 4, Colorado State University, Fort Collins, Colorado.

34. Yevjevich, V. M., 1972, Probability and Statistics in Hydrology Water Resources Publications, Fort Collins, Colorado.

35. Young, G. K. and Pisano, W. C., 1968, Operational Hydrology Using Residuals, Proc. ASCE, Vol. 94, No. HY4, pp. 909-923.

## CHAPTER 14

## RISK

### The Need for Risk Analysis

The most important question facing the designer of any hydrologic structure is: what is the risk of failure? The price of failure of a major dam is high and the risk of this occurrence must be minimized. A study of over 1600 dams (in (8)) has shown the following causes of failure:

| | |
|---|---|
| Foundation problems | 40% |
| Inadequate spillway | 23% |
| Poor construction | 12% |
| Uneven settlement | 10% |
| High pore pressure | 5% |
| Acts of war | 3% |
| Embankment slips | 2% |
| Defective materials | 2% |
| Incorrect operation | 2% |
| Earthquakes | 1% |

In a more recent study of over 300 dam disasters (8) it was found that roughly 35% of the failures were due to inadequate spillway design. Also of importance here is the study of dam failures noted by the AWWA (3). Inadequate spillway design is usually caused by inadequate design flood analysis and this is the direct concern of the hydrologist. Design floods are estimated either from frequency techniques or as the Probable Maximum Flood.

As noted in Chapter 1, the technique of Probable Maximum Flood, despite its name, is a totally deterministic concept and as such has no risk associated with it. Because there is no proof of the existence of extreme boundaries in the meteorological factors which cause floods (19) the concepts of maximum probable precipitation, maximum probable flood and other similarly named imaginary events may be considered as arbitrary. They are concepts of expediency.

Frequency analysis, on the other hand, accepts events of any magnitude as being possible although as the magnitude increases so the probability of occurrence decreases.

The simplest procedure in the frequency analysis estimation of spillway design floods is to select a return period and use either graphical techniques or a mathematical distribution to derive the corresponding event magnitude. Some of the return periods commonly used for different types of structure are (in (7)):

Major dams with probable loss of life

| | |
|---|---|
| Earth dam | 1000 years |
| Masonry or concrete dam | 500 years |
| Costly dams with no likelihood of loss of life | 500 years |
| Moderately costly dams | 100 years |
| Minor dams | 20 years |

In addition, McCaig and Erickson (12) note that in the past it has been common practice to design major dams for floods having theoretical return periods of up to 10,000 years. The ASCE Hydraulics Division Committee on Hydrometeorology (2) has suggested that the Probable Maximum Flood is perhaps equivalent to a design return period of 10,000 years. This elementary procedure takes no account of the increase of risk with increasing project life or of the economically optimum design.

## Economic Design

A second procedure sometimes used in the design of hydraulic structures relates the design spillway capacity not only to the magnitude and frequency of possible floods but also to the monetary value of the dam, the unit cost of the spillway and the value placed upon the lives and property of the people downstream of the dam. McCaig and Erickson (12) have provided a very clear description of this method of design using in their example log-normal distributions of fall and spring floods.

If the average annual losses for a particular structure can be expressed as:

$$C_1 = \Sigma\Delta L\ P \tag{14-1}$$

where $\Delta L$ is the incremental average loss for a particular design flood, x, in dollars and P is the exceedence probability of that design flood; and if the average annual cost of the spillway is given by:

$$C_2 = \Delta x\ Q \tag{14-2}$$

where $\Delta x$ is the incremental cost, in dollars per cfs, of providing spillway capacity for flow Q crs; then the optimum structure design will occur when

$$C = C_1 + C_2 \tag{14-3}$$

is at a minimum. That is to say, for a particular structure and a set of flood flows there will result a particular value of C. By repeating the same set of flood flows with different structure capacities a graph of C versus capacity or design flood can be obtained.

McCaig and Erickson (12) assumed a 2-parameter lognormal probability distribution for flood events so that:

$$p = \frac{1}{\sqrt{2\pi}\,\sigma_y} \int_y^{\infty} e^{-\frac{(y-\mu_y)^2}{2\sigma_y^2}}\, dy \qquad (14\text{-}4)$$

where y is the logarithm of the flood event, x, and $\mu_y$ and $\sigma_y$ are respectively the population mean and standard deviation estimated from the logarithms of the recorded flood events. Substituting Equations 14.4, 14.1 and 14.2 into Equation 14.3, differentiating and equating to zero, the optimum design capacity, $Q_d$, can be obtained.

The ASCE (2) has recently described a similar procedure to McCaig and Erickson but designed for the re-evaluation of the spillway capacity of existing dams. A series of alternate project designs are identified by their spillway design floods, e.g. the 500 year design project, the 1000 year design project, etc.. This series would include the existing project. For each of the possible projects the costs associated with an array of floods with return periods varying from very low to very high are determined.

Damages caused by the various floods to each of the alternate project designs should include upstream damages (in the event of overtopping and subsequent failure of the dam) to recreation, piers, boats, buildings, loss of power, loss of water supply; to the structure itself including dam fill eroded, repair time, powerhouse losses, switchyard losses, etc., and damage downstream of the dam including deaths, injuries, property damage, compensation for loss of water supply, power supply, telephone, road access and lost employment. It is instructive to note that in the ASCE example (2) death was valued at $150,000, permanent disabling injury at $200,000 and a non-disabling injury at $10,000. The property damages should be determined by carrying out a stage-damage analysis using measured flood profiles.

For each project the average annual risk can be calculated by arithmetic strip integration of the area beneath the return period-damage curve. The cost of each alternate project design is known and can be converted to an average annual cost. This cost, sometimes known as the "operating rate" (12), may include items for interest, taxes, depreciation, etc., and normally ranges between 8 and 10 percent of the total capital cost. Curves of the type shown in Figure 14-1 can then be drawn and the optimum project design determined.

Note that the series of alternate projects might consist of one dam design with floods of successively longer return

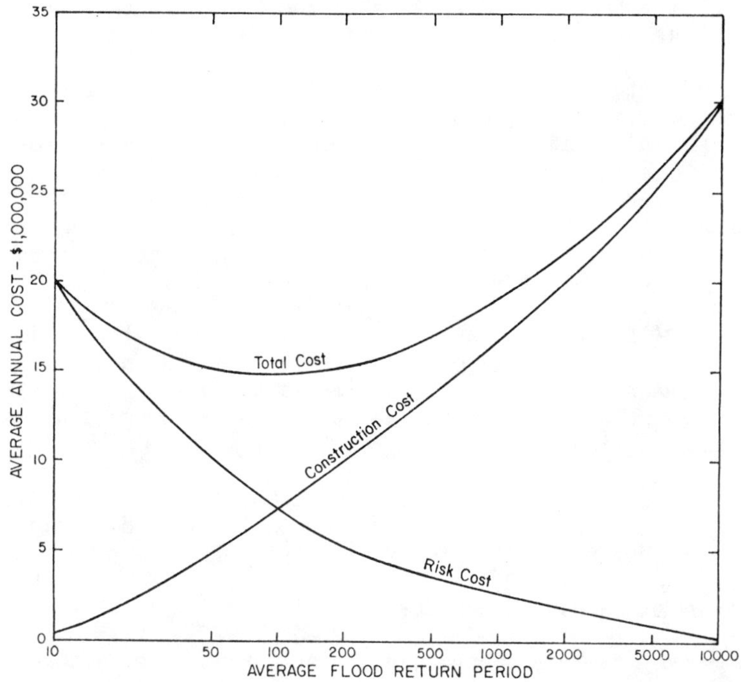


Figure 14-1. Average annual costs for different designs (example only).

period being accommodated by a longer spillway, by downstream flood protection work, by paving the dam top and downstream dam surface to reduce erosion from overtopping, by construction of an upstream reservoir to reduce inflows or by other similar means.

Risk Design

Neither of the two techniques described so far include the concept of total risk. For any hydraulic structure there is a total risk of failure which can be broken down into the risk of failure of each project component i.e. hydrologic, hydraulic and structural. The risk within any component can then be broken down into true risk and uncertainty. Yen and Ang (18) have used the terms objective risk and subjective risk.

For the hydrologic component, risk is the calculable probability of failure e.g. occurrence of a certain flood, occurrence of a drought, etc.. The calculation of risk is based on the assumption that the underlying event distribution is known. As an example, if it is known that flood magnitudes in a particular river valley location follow the lognormal distribution and that the time-distribution of the floods follow a Poisson distribution then the risk that the flood of a certain magnitude will occur in the next five years can be computed exactly.

Uncertainty occurs because the basic data available contains random measurement and computation errors, systematic errors, non-homogeneity in time, loss of information in changing from a continuous record to a discrete data set and so on. These imperfect data are then used to estimate the parameters of the assumed population distribution. Uncertainty generally increases as the variance of the sample data increases and decreases as the sample length increases.

Thomas (16) has evaluated the errors in streamflow estimates made from a continuous stage record while Moss (13) has related the standard error of discharge estimates to the number of streamflow measurements made per year and the associated costs of maintaining the station.

The effect of uncertainty on the parameters of the population distribution can be included in an analysis by computing the standard error of estimate of the particular distribution at the required probability level. Confidence limits around the expected event magnitude can then be calculated.

To summarize this concept, hydrologic risk is made up of basic risk and uncertainty both of which can be evaluated. What cannot be evaluated is the error caused by selecting the wrong distribution to fit the sample data. It is true that the goodness of fit of a distribution, once chosen, can be measured using the Chi-Square or Kolmogorov-Smirnov or similar tests and thus the best-fitting distribution can be selected. Generally, however, the sample data will occupy the central portion of a frequency distribution while the event magnitudes which it is required to compute will be in the extremes so that the best-fitting distribution may not necessarily be the best to use.

The computation of standard errors of estimates for various common distributions has been described earlier. The remainder of this chapter will cover the calculation of basic risk, the assumption being made that the underlying distribution is known.

Suppose that for a time invariant hydrolgic system the probability of occurrence of an event, x, greater than the design event, $x_o$, during a period of n years is P. Then the probability of non-occurrence, Q, is 1-P.

If this design event has a return period of T-years and a corresponding annual probability of exceedence of p then:

$$p = \frac{1}{T} \tag{14-5}$$

the probability of non-occurrence in any one year is:

$$q = 1 - \frac{1}{T} \qquad (14\text{-}6)$$

the probability of non-occurrence in n years is:

$$Q = (1 - \frac{1}{T})^n \qquad (14\text{-}7)$$

So that, finally, the probability that x will occur at least once in the n years is:

$$P = 1 - (1 - \frac{1}{T})^n \qquad (14\text{-}8)$$

This is the risk of failure and is based on the assumption of independence of annual events. Yen (17) has tabulated values of T, the required design return period, for various expected project lives, n, and permissible risks of failure, P. Table 14-1 is adapted from Yen. Figure 14-2 is based on the solution of Equation 14-8.

Table 14-1

Design Return Period for Various Project Lives and Risks of Failure[1]

| Permissible risk of failure | Expected Project Life, n, in years | | | | | | | |
|---|---|---|---|---|---|---|---|---|
| | 1 | 2 | 5 | 10 | 20 | 25 | 50 | 100 |
| 0.99 | 1.01 | 1.11 | 1.66 | 2.71 | 4.86 | 5.95 | 11.4 | 22.2 |
| 0.95 | 1.05 | 1.29 | 2.22 | 3.86 | 7.16 | 8.85 | 17.2 | 33.9 |
| 0.90 | 1.11 | 1.46 | 2.71 | 4.86 | 9.19 | 11.4 | 22.2 | 43.9 |
| 0.75 | 1.33 | 2.00 | 4.13 | 7.73 | 14.9 | 18.6 | 36.6 | 72.6 |
| 0.50 | 2.00 | 3.41 | 7.73 | 14.9 | 29.4 | 36.6 | 72.6 | 145. |
| 0.33 | 3.00 | 5.45 | 12.9 | 25.2 | 49.9 | 62.1 | 124. | 247. |
| 0.25 | 4.00 | 7.46 | 17.9 | 35.3 | 70.0 | 87.3 | 174. | 348. |
| 0.20 | 5.00 | 9.47 | 22.9 | 45.3 | 90.1 | 113. | 225. | 449. |
| 0.10 | 10.0 | 19.5 | 48.0 | 95.4 | 190. | 238. | 475. | 950. |
| 0.05 | 20.0 | 39.5 | 98.0 | 195. | 390. | 488. | 975. | 1,950. |
| 0.02 | 50.0 | 99.0 | 248. | 495. | 990. | 1,238. | 2,476. | 4,951. |
| 0.01 | 100. | 199.5 | 498. | 995. | 1,990. | 2,488. | 4,977. | 9,953. |

[1] From Yen (17)

If the series of recorded or measured events are not an annual series but a partial duration series with an average of K observations per year then the probability that the T-year event will be equalled or exceeded in n consecutive years is:

$$P = 1 - [1 - 1/T\ K]^{nK} \qquad (14\text{-}9)$$

Figures similar to Figure 14-2 (for which K = 1) can be drawn for all values of K, (1).

If failure is associated not with exceedence of the design event but with failure to reach the design event, e.g. a drought design, then the return period must be redefined.

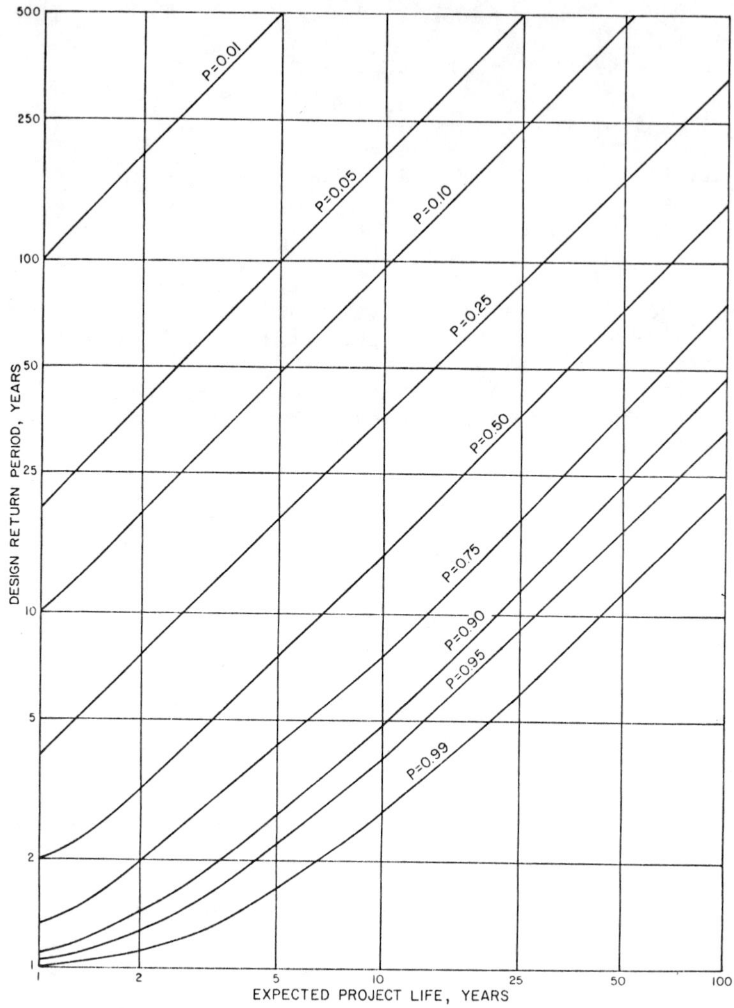


Figure 14-2. Theoretical probability of failure for given project life and design return period.

In the event that the design return period is made equal to the expected project life there is a 63.4% chance of failure of the project. This can be shown in Equation 14-8 by putting T = n:

$$P = 1 - (1 - \frac{1}{n})^n \tag{14-10}$$

In the limit as $n \rightarrow \infty$

$$(1 - \frac{1}{n})^n \rightarrow \frac{1}{e} = 0.368 \tag{14-11}$$

and so, for large n, P tends to 63%.

Similarly, supposing that a project has been designed against a hydrologic event of return period T years then the risk of failure after completion of n' years of the expected project life of n years can be calculated (14).

Writing Equation 14-7 as

$$Q = [(1 - \frac{1}{T})^T]^{T/n} \qquad (14\text{-}12)$$

and using the same asymptotic approximation as in Equation 14-11, Gill (10) has shown that for a given value of P or Q there is a linear relationship between T and n, as:

$$Q = (\frac{1}{e})^{T/n} \qquad (14\text{-}13)$$

$$n = T \ln (1/Q) \qquad (14\text{-}14)$$

A frequently used approximation resulting from Equation 14-8 is:

$$T \sim n/P \qquad (14\text{-}15)$$

Gumbel (11) termed this the "design quotient".

The probabilities referred to above are all probabilities of occurrence of an event of a certain magnitude. Also of interest is the average probability of occurrence of all events above that certain magnitude. For example, in a series of n annual events the number, m, of events which equal or exceed the T-year event in (n+1)/T. The annual probability of occurrence of the maximum event is 1/(n+1), of the second largest event is 2/(n+1), of the third largest event is 3/(n+1), etc. so that, the average probability $\bar{p}$ of the n' events which exceed the T-year event is given by:

$$\bar{p} = (\frac{1}{n+1} + \frac{2}{n+1} + \frac{3}{n+1} + \ldots + \frac{n'}{n+1})/n' \qquad (14\text{-}16)$$

Benson (6) has shown that this expression reduces to:

$$\bar{p} = (n+T)/2T(n+1) \qquad (14\text{-}17)$$

which, as n approaches infinity, becomes

$$\bar{p} \sim 1/2T \qquad (14\text{-}18)$$

Thus, in general, the average probability of occurrence of all events above the T-year event is approximated by the probability of the 2T-year event. For example, the average probability of occurrence of all events greater than the 100-year event is approximately 0.005, which corresponds to the 200-year event.

The expressions developed so far in this chapter have all been distribution-free, that is, no assumptions have been made regarding the underlying event distribution. If it is required to estimate the event magnitude corresponding to the design return period computed from, for example, Equation 14-8 then a probability distribution must be assumed.

To show the wide variation possible in the results of this assumption of a distribution, Figure 14-3 is reproduced from Gumbel (11). This figure shows the relationship between design quotient and the reduced variable, z, where

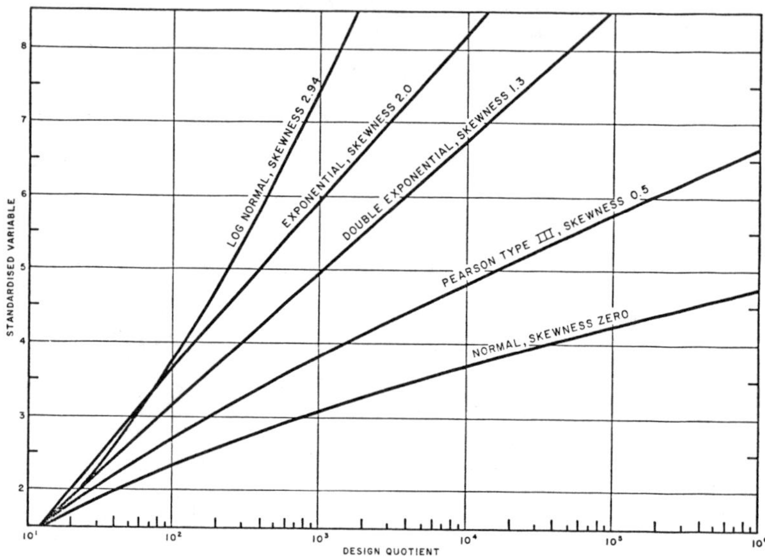


Figure 14-3. Standardized variable vs design quotient.

$$Z = (x - \mu)/\sigma \qquad (14\text{-}19)$$

for several commonly used distributions.

Those distributions shown on Figure 14-3 which have fixed coefficient of skew are the normal with $\gamma_1 = 0$ and the double exponential (or type I extremal) with $\gamma_1 = 1.3$. The coefficients of skew have been arbitrarily chosen for the other distributions shown on the figure.

Alternatively, if the assumptions are made that events are independent and the mean number of events in unit time is constant then the Binomial and Poisson distributions can be used to evaluate risk, as described in Chapter 4. For a Poisson distribution of event occurrences and an extremal type I distribution of event magnitudes, Shane (15) has defined the design event, x, as:

$$x = v + \gamma \ln (\lambda F) \tag{14-20}$$

where v is a base flow, $\gamma$ is a parameter of the extremal distribution, $\lambda$ is the expected rate of occurrence of events, $\lambda = np$, in the Poisson process and F is the risk factor. The maximum likelihood estimates of $\gamma$ and $\lambda$ are given (15) as:

$$\hat{\gamma} = \bar{x} - v \tag{14-21}$$

and

$$\hat{\lambda} = n_e/n \tag{14-22}$$

where $n_e$ is the number of events recorded and n is the period of record.

Benson (5) investigated the variation which occurs when small samples are used to estimate a frequency distribution for which the parameters are known exactly. Starting with a known frequency curve Benson (5) constructed short random data sets, drew best-fitting curves and estimated events at different return periods from those curves. From a basic set of 1000 points, 100 records of ten points, 40 records of 25 points, 20 records of 50 points and 10 records of 100 points each were drawn. It was found that records of up to 25 points could not define satisfactorily even short-term events. Long-term records (40 to 50 points) were found to define events magnitudes up to the length of records with reasonable accuracy.

Yen and Ang (18) have described a procedure for designing hydraulic structures on the basis of a risk analysis. Using as an example the design of an urban sewer system, an overall project risk was chosen on the basis of possible property damage. The hydraulic and hydrologic risks are combined as $\alpha_h$ and are related to the structural risk, $\alpha_s$, and overall risk, $\alpha$

$$(1 - \alpha) = (1 - \alpha_s)(1 - \alpha_h) \tag{14-23}$$

Yen and Ang then defined the combined hydraulic and hydrologic risks as:

$$\alpha_h = P(x > Q_c)\ P(N > \nu) \tag{14-24}$$

where $P(x > Q_c)$ is the probability of an event X exceeding a design event, $Q_c$, (the hydrologic risk) and $P(N > \nu)$ is the probability that N, a random variable, will exceed $\nu$, a safety factor (hydraulic risk), where

$$\nu = Q_b/Q_c \tag{14-25}$$

and $Q_b$ is the discharge actually used in design. N was assumed to be distributed lognormally with unit mean and a variance, $\sigma_N^2$, equal to the total of the variances of the uncertainties such as inaccuracy of measurement, systematic errors in computation, etc.

$$\sigma_N^2 = \sigma_1^2 + \sigma_2^2 + \ldots \tag{14-26}$$

as discussed earlier in the chapter.

If $\alpha$ and $\alpha_s$ are known, then $\alpha_h$ can be determined from Equation 14-23 and for various values of $\nu$, the safety factor, corresponding values of $P(X > Q_c)$, the hydrologic risk, can be found. The equivalent design return period can be found from Equation 14-8 and, assuming a probability distribution to fit the observed data, the corresponding event magnitude, $Q_c$, is found. Yen and Ang (18) used a type I extremal distribution although any other suitable distribution could equally well have been used. By plotting values of $Q_c$ versus $\alpha$ (or $Q_b = \alpha\ Q_c$ vs $\alpha$) the optimum discharge can be found. Thus, by defining rigorously the hydrologic risk, the common hydraulic practice of using a safety factor to include the effects of hydraulic risk is provided with a scientific basis.

In the event that no streamflow records are available at the design site, Davis et al. (9) have described a method of evaluating uncertainty by considering the distribution of rainfall events. If the number of rainfall events per season, N, is Poisson distributed with mean $\lambda$, i.e.

$$P_N(x|\lambda) = \frac{\lambda^x e^{-\lambda}}{x!} \tag{14-27}$$

and if the amount of rainfall, R, per event is exponentially distributed

$$P_R(k|u) = ue^{-uk} \tag{14-28}$$

where $1/u$ is the mean rainfall per event, then the return period of k units of rain in a season, T, is

$$T_R(k|\lambda,u) = [1 - \exp(-\lambda e^{-uk})] \tag{14-29}$$

By using a linear rainfall-runoff relationship

$$Q = C\ (R-A) \tag{14-30}$$

where C is a coefficient depending upon the rainfall characteristics of a given watershed and A is a measure of initial abstractions, also depending on the watershed, then an expression for the probability density distribution of the flood return period, $T_Q$, can be given as

$$T_Q(y|\lambda,u) = [1 - \exp\{-\lambda + P_Q(y|u)\}]^{-1} \qquad (14\text{-}31)$$

where $P_Q(y|u)$ is the distribution function of runoff per event.

Uncertainty is included in the analysis (9) be considering the parameters λ, u and c as variables. Davis et al. assumed that λ and u could be described by a 2-parameter gamma distribution while a beta distribution was used for c.

The results of this approach provide design flows relying only on rainfall data for watersheds with ungauged streams by taking into account the uncertainty of the site parameters. It was found (9) that a closed form solution was not possible and so data generation was used to derive the distribution of the flood return period.

To conclude this chapter on hydrologic risk, a final note on the accumulation of risk. On an individual basis a design return period of 1000 years is often considered safe. When it is considered, however, that there are now approximately 10,000 large dams in the world (7), 1000 of which can be thought of as in independent basins, then since 0.001 x 1000 = 1.0, the 1000-year event will be equalled or exceeded each year at at least one of the dam sites. Similarly, Alexander (1) has shown that in Japan, where there are about 1700 dams, the average design return period for spillway design floods is of the order of 200 years. It would be expected, therefore, that on the average 8 or 9 dams will incur design floods annually.

## References


1. Alexander, G. N., 1969, Application of Probability to Spillway Design Flood Estimation, Proc. IASH Symposium on Floods and their Computations, Leningrad, 1967, IASH-UNESCO-WMO Studies and Reports in Hydrology No. 3, pp. 536-543.

2. ASCE, 1973, Re-evaluating Spillway Adequacy of Existing Dams, Report of the Task Committee on the Re-evaluation of the Adequacy of Spillways of Existing Dams of the Committee on Hydrometeorology of the Hydraulics Division, Proc. ASCE, Vol. 99, No. HY2, pp. 337-372.

3. American Water Works Association, 1966, Spillway Design Practice, AWWA Manual No. M13, New York.

4. Banerji, S. and D. K. Gupta, 1969, On a General Theory of Duration Curve and its Application to Evaluate the Plotting Position of Maximum Probable Precipitation or Discharge, Proc. IASH Symposium on Floods and their Computations, Leningrad, 1967, IASH-UNESCO-WMO Studies and Reports on Hydrology No. 3, pp. 183-193.

5. Benson, M. A., 1960, Characteristics of Frequency Curves Based on a Theoretical 1000-Year Record, USGS Water Supply Paper No. 1543-A, pp. 51-73.

6. Benson, M. A., 1967, Average Probability of Extreme Events, Water Resources Research, Vol. 3, No. 1, p. 225.

7. Biswas, A. K., 1971, Some Thoughts on Estimating Spillway Design Flood, Bull. IASH, Vol. XVI, No. 4, pp. 63-72.

8. Biswas, A. K. and S. Chatterjee, 1971, Dam Disasters: An Assessment, J. Engineering Institute of Canada, Vol. 54, No. 3, pp. 3-8.

9. Davis, D. R., L. Duckstein, C. C. Kisiel and M. M. Fogel, 1973, A Decision - Theoretic Approach to Uncertainty in the Return Period of Maximum Flow Volumes Using Rainfall Data, Proc. UNESCO-WMO-IASH Symposium on the Design of Water Resources Projects with Inadequate Data, Madrid, Vol. 1, pp. 63-74.

10. Gill, M. A., 1972, Analysis of Probability and Risk Equations, Proc. ASCE, Vol. 98, No. HY5, pp. 969-971.

11. Gumbel, E. J., 1955, The Calculated Risk in Flood Control, Appl. Sci. Res., Sec. A, Vol. 5, pp. 273-280.

12. McCaig, I. W. and O. M. Erickson, 1959, Spillway Capacity and Flood Flows, Proc. Symposium No. 1, Spillway Design Floods, NRC, Ottawa, pp. 262-287.

13. Moss, M. E., 1969, Maximization of Net Benefit from a Streamgauge, Proc. Fiftieth Annual Meeting of the American Geophysical Union, Washington, D. C.

14. Prasad, T., 1971, Discussion of "Risks in Hydrologic Design of Engineering Projects", Proc. ASCE, Vol. 97, No. HY1, pp. 201-202.

15. Shane, R., 1966, A Statistical Analysis of Base-Flow Flood Discharge, PhD Thesis, Cornell University.

16. Thomas, R. B., 1971, Errors in Streamflow Estimates from Continuous Stage Records, Proc. Symposium on Statistical Hydrology, Tucson, Arizona.

17. Yen, B. C., 1971, Risks in Hydrologic Design of Engineering Projects, Proc. ASCE, Vol. 96, No. HY4, pp. 959-966.

18. Yen, B. C. and A. H. A. Ang, 1971, Risk Analysis in Design of Hydraulic Projects, Proc. Symp. on Stochastic Hydraulics, Univ. Pittsburgh, pp. 694-709.

19. Yevjevich, V., 1968, Misconceptions in Hydrology and their Consequences, Water Resources Research, Vol. 4, No. 2, pp. 225-232.

## CHAPTER 15

## SUMMARY AND CONCLUSIONS

### Introduction

The magnitude and frequency of occurrence of extreme hydrologic events is of every day importance in most parts of the world. Since man has, for reasons of communication, water supply, agriculture, etc., built most of his communities on the flood plains of large rivers his life-style is extremely susceptible to flood damage. Today's pressure of population increases the density of development along the river banks. The flood of 1948 on the Fraser River in British Columbia caused $20 million of damage. It has been estimated that if the same magnitude of flood occurred today the damage would be over $200 million.

At the opposite end of the water spectrum, the production of sufficient food to feed the world's rapidly increasing population necessitates the increasing use of irrigation. Mankind thus becomes ever more susceptible to disaster through drought.

Proper use of existing hydrologic techniques could, through flood plain zoning and efficient design techniques, eliminate much of the present loss of life and damage caused by floods and droughts.

In the 1920's and 1930's the introduction of simple statistical analysis gave an impetus to the science of hydrology. Very soon, however, a general distrust of probability methods began to grow because too many users knew too little about statistical analysis and they apparently expected the methods to overcome the lack of data (9).

A growing use of deterministic methods, replacing the fall in popularity of probability techniques, led to the development in the late 1930's of the unit hydrograph principle. Advances in meteorology enabled the conditions producing storm rainfall to be analysed, with the result that maximum rain producing storms could be synthesized. This technique produced a large number of new, rather vague, technical terms such as Probable Maximum Precipitation, Maximum Possible Precipitation, Standard Project Storm, Maximum Probable Flood, etc., based on the premise that some definite limit existed for all the variables responsible for flood events and that, subsequently, some limit must apply to the flood runoff itself. The drawback to this method is that no probability level can be assigned to the "probable" events, because of their deterministic origins. Similarly no confidence limits can be applied to these events and the non-specialist is left with the impression that these estimates are 100% accurate with no risk involved.

The philosophical error in the Maximum Probable argument has been described by Yevjevich (27) and others. It is not reasonable to say that a precipitation of 30 inches in one hour can occur but a precipitation of 30.1 inches cannot. The probability approach states that any variable has a finite probability of reaching any value between zero and infinity. As the variable increases in magnitude so the associated probability of occurrence decreases and for very large events approaches zero. However, given enough time, even the improbable becomes certain. The essential stochasticity of precipitation has been recognized by Yevjevich (28) as being mainly due to the random nature of atmospheric variables such as opacity and transmitted radiation. The stochastic precipitation events are then somewhat attenuated by the water and energy storages of the oceans and continents to produce runoff events of mixed deterministic/stochastic nature.

Today, the necessity of producing economically designed projects has produced a need for hydrologic risk analysis and a corresponding upsurge in the use of probabilistic methods in hydrology. Compared to the 1920's and 1930's, however, more data is available and the theory of sampling errors and risk analysis is better understood. Problems still exist in the statistical techniques, however, and a certain amount of subjectivity is still involved, particularly in the choice of a frequency distribution to fit to the observed data.

## Data Abstraction, Graphs, and Plotting Positions

Data for frequency analyses may be abstracted from the recorded data using either annual series or partial duration series. Annual series consist of one event per year; partial duration series consist of all events above a base magnitude, regardless of time of occurrence. The partial-duration series method would initially seem advantageous in that more data, and hence information, are incorporated. However, these additional data increase the definition of event magnitudes only in the central part of the frequency curve which is the area of least interest. Use of the partial-duration series always involves the arbitrary establishment of a base flow and sometimes requires subjective decisions regarding the independence of adjacent events. For these reasons and because the annual series is simpler to abstract and analyse it is to be preferred in frequency studies.

Riggs (20) and Benson (2) have detailed many other reasons, but, to summarize, it is usually stated that mathematical fitting of a standard probability distribution is preferable to plotting a graph and fitting a curve by eye, because:

(a) mathematical fitting is theoretically better, and
(b) mathematical fitting eliminates the subjectivity of individual judgement.

In fact, curve fitting by eye and by probability function are both empirical since the true distribution of the recorded events is not known. In addition the very lack of subjectivity of the mathematical procedure is sometimes a disadvantage; the inclusion or exclusion of one or two events may result in large changes in the resulting frequency relationship. The mathematical procedure incorporates all data whereas the individual drawing in a curve by eye may elect to ignore some events in order to get a better-fitting curve. A method is not better simply because it leads to uniform answers, if those answers are uniformly unsound (1).

An intermediary or semi-graphical method exists in which graphs are used to fit curves of standard probability distributions to the data. The procedure recommended is to use a mathematical method of fitting a standard probability distribution and to arrange for the output of the method always to include a plot of the data points and the fitted curve. In this way the procedure can be standardized and automated for machine computation while retaining the option of looking at a graph and reviewing the fitted line on the basis of engineering experience.

Any type of plot of extreme events requires the consideration of plotting positions. When all analysis computations were done by hand, graphical techniques of analysis were extremely attractive because of their brevity and simplicity and the choice of plotting positions was therefore of great importance. Today, when all calculations are performed by computer, graphs are used only as a pictorial form of output presentation and the problem of plotting position has faded somewhat in importance. With this in mind it would be difficult to justify use of any plotting position other than the mean frequency

$$p = m/(n+1) \qquad (15\text{-}1)$$

where m is the order of the event in the sorted series of n observed events. For the largest event in the series m = 1 and for the smallest event in the series, m = n. This method gives conservative results in that the return period conforms closely to the period of record. Benson (3) has demonstrated that this is the best plotting position to use for economic studies of hydrologic design.

## Frequency Distributions

*Selection of a Distribution* - The primary objectives of frequency analysis are to determine the return periods of recorded events of known magnitude and then to estimate the magnitude of events for design return periods beyond the recorded range. The intermediary between these two objectives is the theoretical probability distribution. The sample data are used as an estimate of an unknown population to calculate the parameters of the selected probability distribution. The fitted distribution is then used to estimate event magnitudes corresponding to return periods greater than or less than those of the recorded events.

There is no general agreement among hydrologists as to which of the various theoretical distributions available should be used. The present state of the art is also such (4) that no general agreement has been reached as to preferable techniques, and no standards have been established for design purposes. As examples of this divergence of choice, Spence (24) compared the fit of the normal, 2-parameter lognormal, type I extremal and log-type I extremal distributions to annual maximum flows on the Canadian Prairies and found that the lognormal was the best fitting; Cruff and Rantz (10) compared six probability distributions in California and found that the Pearson type III was the most desirable. In other studies, Santos (21) has found the lognormal distribution better than the Pearson type III, Gumbel (II) has explained that "It seems that the rivers know the [extreme value] theory. It only remains to convince the engineers...of the validity of this analysis", and Benson (2) has found in a study of 100 long term flood records that no one type of frequency distribution gives consistently better results. In the United States, Reich (19) conducted a survey of engineers and hydrologists and found that of the extremal type I, log extremal type I and log-Pearson type III, the log-Pearson type III was preferred. In Italy, Cicioni et al. (8) tested the 2-parameter lognormal, 3-parameter lognormal, 2-parameter gamma, Pearson type III and extremal type I distributions on 108 data sets and found the 2-parameter lognormal to be the most suitable.

In short, no one distribution is acceptable to all hydrologists.

Acceptance of a certain distribution for analysis of flood peaks must be based on the goals and conditions that are to be fulfilled and satisfied by the distribution (29). Goodness of fit is a necessary but not a sifficient condition for acceptance. Goodness of fit tests often used include Chi-Square and Kolmogorov-Smirnov (26). If goodness of fit were the only criterion, then high order polynomials would often provide a much better fit than any of the standard distributions, and yet this method is not used because there is no hydrologic justification. The most important criteria in the selection of a model are that there be a sound theory describing the phenomenon and that the model should abstract the maximum information from the data using proper estimation techniques.

This was realized by the recent U. S. Water Resources Council Work Group (4) who wrote that "no single method of testing the computed results against the original data was acceptable to all those on the Work Group, and the statistical consultants could not offer a mathematically rigorous method." The Work Group concluded that a frequency distribution could not be chosen solely on statistical grounds but recommended that the log-Pearson type III distribution be used because as a 3-parameter distribution it offers considerable flexibility, for a zero skewness it reduced to the lognormal distribution, and finally because it is in common use by U. S. Government agencies.

The problem of choosing a distribution is not restricted to hydrology by any means. In the field of biosciences, Katti and Sly (12) summarized their findings as:

(a) No single theoretical distribution has been found to describe any large scale data.

(b) For a number of data there could be two or more theoretical distributions that fit equally well and there is no way to choose between them based on fit alone.

(c) Two or more physical models could lead to the same final statistical distribution and hence the estimation of the parameters of the distribution may not have unique meaning.

(d) "...Different methods of estimation lead to widely differing estimates when the methods are consistent...there are a number of empirical frequencies to which the same theoretical frequency function has been fitted by different consistent methods...."

Although statistical methods cannot by themselves determine the correct frequency distribution, they can, in some cases, provide reasons why distributions may not be suitable. As an example some distributions such as the type III Extremal, Pearson Type III and log-Pearson type III require the estimation of the coefficient of skewness from the sample data. It is well known that the variability of sample estimates of the coefficient of skew is large (10) and this may be sufficient reason to prefer some other distribution. On the other hand use of 2-parameter distributions implies a fixed value of the coefecient of skew which may not be valid either.

As a second example of this process of elimination the following objections have been raised (2), to the use of the type I extremal distribution (Gumbel) for flood flows:

(a) It is assumed that the treatment derived for daily discharges can also be applied to instantaneous flows.

(b) The daily discharges are not independent events.

(c) The 365 daily discharges in a year do not constitute a large number as required by the theory of extreme values.

(d) An assumption underlying the extreme value theory is that all the events are part of the same statistical population. Yet, in many cases, the annual maximum event may be due to a variety of causes such as normal rainfall, snowmelt or hurricane. There are different physical factors controlling each of these types of events. The assumption of one population therefore may not be valid. Whether or not these objections to the use of the type I extremal distribution are sufficiently serious to deter use, is a matter of opinion.

Other statistical restrictions on the use of certain distributions which may help in choosing one distribution from many include:

(a) The truncated normal distribution (with the probability of events being less than or equal to zero being replaced by a probability mass, P, at $x = 0$) may be used only if the sample coefficient of skew is very close to zero and the sample data are such that P is very small.

(b) The 2-parameter and 3-parameter lognormal distributions may be satisfactory only if the coefficients of skew of the reduced data ($\ln x$ or $\ln x-a$) are very close to zero.

(c) The type I extremal distribution will be suitable only when the coefficient of skew of the recorded events is very close to 1.13.

(d) The Pearson type III distribution is only unbounded at the upper end for positive coefficient of skew.

(e) The log-Pearson type III distribution is only unbounded at the upper end for positive coefficients of skew of the logarithms and when the parameters $\beta$ and $1/\alpha$ are greater than zero.

Of lower importance than the theoretical background is the ease of computation. Matalas and Wallis (15) found several computational difficulties in obtaining maximum likelihood solutions for the Pearson type III and log-Pearson type III distributions and recommended that the use of other distributions warrants consideration. Similarly Pentland and Cuthbert (18) found that use of the log-Pearson type III distribution for the Fraser River Basin, in British Columbia, led to large discontinuities and unnatural flood frequencies. They substituted the lognormal distribution in place of the log-Pearson type III.

*Estimation of Parameters* - There are four main methods of estimating distribution parameters: moments, maximum likelihood, least squares and graphical. It is often accepted (15) that the method of maximum likelihood is the most efficient method and should be used wherever possible. By computing relative efficiencies, Matalas (14) has shown that for low flow analysis the method of moments uses only one-half of the sample information extracted by maximum likelihood for a shape parameter of about 10.

*Frequency Factors* - For any distribution the T-year event magnitude can be computed (7) from a general equation of the form:

$$x(K) = \mu + K\sigma \tag{15-2}$$

where $\mu$ and $\sigma$ are sample estimates of the population mean and standard deviation and K is a frequency factor specific to the chosen distribution. For each probability distribution the frequency factor K, can be derived from the same size, sample parameters, etc.

*Confidence Limits* - Once a suitable probability distribution has been chosen and the magnitude of the event, x(K), at the required return period, T, has been computed from the general frequency equation then upper and lower confidence limits should be established for this event magnitude.

From the frequency curve, two methods are available to compute confidence limits. The analytical method uses numerical techniques to integrate the probability density function of the sample quantile. For practical distributions, however, this is very difficult and an empirical method is often used instead. The empirical method computes the standard error of x(K), s(K), and then assumes that the T-year event is normally distributed with mean x(K) and standard deviation s(K) so that the confidence interval is

$$x(K) \pm t\ s(K) \tag{15-3}$$

where t is the standard normal deviate at the required confidence level.

The validity of the empirical method rests on the assumption of normality of the distribution of T-year events. This has been tested by data generation and the results have been described (13).

## Regional Analysis

Regional analysis techniques provide a means of combining records from many gauges. This provides the two advantages of reducing standard errors of estimates at gauge sites and enabling estimates to be prepared for ungauged sites.

The description of regional analysis techniques given in Chapter 13 divided the techniques into five main methods: index flood, multiple regression, square grid, modified single station probability distributions, and regional record maxima. Each of the first four methods contains several variations, however, and these variations tend to overlap between methods so that the division becomes somewhat artificial. The index-flood method, as originally proposed and put into practice suffers from many defects. The most serious defects are that the method is totally empirical, the distribution of peak events is not known and all relationships are derived by graphical curve-fitting.

The square grid method has a logical ordering of basic data which, when combined with the convenience of automatic data processing, provides a data file of regularly spaced physiographic, meteorologic and hydrologic data easily and rapidly accessible for many types of study. The data file can be regularly updated and may be enlarged by the addition of more parameters as they become available. Using this data file as a base, an extremely versatile series of analyses becomes possible. Originally the square grid method was used to define the areal distribution of mean annual runoff for preliminary province-wide water resources studies (22), but additional steps now available include the generation of synthetic monthly flows at ungauged sites (17), and, with the addition of meteorologic forecasts, the use of a parametric model to provide monthly forecasts of streamflow (23), as well as the provision of flood frequency analyses at ungauged sites.

Using the square grid method for frequency analyses combines the use of a standard frequency distribution with a more efficient data base. In recommending this square grid approach for regional frequency analysis the probability distribution to be used should be selected on the basis of the comments and recommendations previously made.

Hydrologic series frequently consist of observations that are dependent upon one another, they are serially correlated. Thus each observation contains some information which has already been contained in previous observations. A serially or autocorrelated time series therefore contains less information than would an equal number of pure random observations. If a time series of a given length is correlated with a shorter time series, then the two-station correlation can be used to increase the information content of the shorter series. If either or both of the time series are autocorrelated then the increase in information will be less than if the time series were pure random. Similarly if a number of gauging stations are used to estimate regional parameters then the information content of those parameters will be a maximum when all the station records are independent and will decrease as the dependence between records increases.

Bayesian analysis (25) offers the possibility of combining information from regional analysis with single station frequency analyses.

## Risk

Most hydraulic structures are designed on the basis of deterministic Probable Maximum Floods for which the risk is totally unknown. For those structures designed on the basis of frequency analysis, the hydrologic risk is generally accounted for by a simple choice of return period for the design flood. For any given return period however, the risk of project failure is proportional to the expected project life.

An alternate procedure is to initially assign an acceptable overall risk to the project. This overall risk should be computed on the basis of the consequences and costs of failure. This total risk can then be subdivided into structural, hydraulic and hydrologic risk and the allowable hydrologic risk can be determined. On the basis of this allowable hydrologic risk and the expected project life the required return period can be found and by assuming a probability distribution the corresponding T-year event magnitude can be calculated. In this manner, Probable Maximum Floods or other deterministic estimates can often be assigned a risk and used in further analysis.

To account for the uncertainties in estimation of distribution parameters it is usual to compute the upper 95% confidence limit for the T-year event and design for that discharge.

It is well to keep in mind, however, that no matter how sophisticated a risk analysis is undertaken, the unexpected tends to occur with remarkable regularity. As an example (5), the Rincon de Bonete hydroelectric project on the Rio Negro in Uraguay was designed for the 1000-year flood of 325,000 cfs on the basis of 27 years of streamflow records during which the peak flow was 135,000 cfs. Fourteen years after construction (1959) as a result of prolonged heavy rainfall a flood peak of 605,000 cfs was measured. Using the original frequency analysis the 1959 flood has a theoretical return period of 500,000 years. Fortunately, the dam held and about one-half of the flood peak was absorbed by the reservoir which rose 15 feet above its designed maximum level.

<u>References</u>

1. American Water Works Association, 1966, Spillway Design Practice, AWWA Manual No. M13, New York.

2. Benson, M. A., 1962, Evolution of Methods for Evaluating the Occurrence of Floods, USGS Water Supply Paper 1580-A.

3. Benson, M. A., 1962, Plotting Positions and the Economics of Engineering Planning, Proc. ASCE, Vol. 88, No. HY6, pp. 57-72.

4. Benson, M. A., 1968, Uniform Flood Frequency Estimating Methods for Federal Agencies, Wat. Res. Res., Vol. 4, No. 5, pp. 891-908.

5. Biswas, A. K., 1971, Some Thoughts on Estimating Spillway Design Floods, Bull. IAHS, Vol. XVI, No. 4, pp. 63-72.

6. Blench, T., 1959, Empirical Methods, Proc. Symposium No. 1, Spillway Design Floods, NRC, Ottawa, pp. 36-48.

7. Chow, V. T., 1954, The Log-Probability Law and Its Engineering Applications, Proc. ASCE, Vol. 80, pp. 1-25.

8. Cicioni, G., G. Giuliano and F. M. Spaziani, 1973, Best Fitting of Probability Functions to a Set of Data for Flood Studies, Proc. Second International Symposium in Hydrology, Floods and Droughts, Water Resources Publications, Fort Collins, Colorado, pp. 304-314.

9. Clark, R. H., 1959, Opening Address, Symposium No. 1, Spillway Design Floods, NRC, Ottawa, pp. 1-3.

10. Cruff, R. W. and S. E. Rantz, 1965, A Comparison of Methods Used in Flood Frequency Studies for Coastal Basins in California, USGS Water Supply Paper 1580-E.

11. Gumbel, E. J., 1966, Extreme Value Analysis of Hydrologic Data, Proc. Hydrology Symposium No. 5, NRC, Ottawa, pp. 147-169.

12. Katti, S. K. and L. E. Sly, 1965, Analysis of Contagious Data Through Behaviouristic Models, in: Classical and Contagious Discrete Distributions, ed. G. P. Patil, Statistical Publishing Society, Calcutta, pp. 303-319.

13. Kite, G. W., 1975, Confidence Limits for Design Events, Wat. Res. Res., Vol. 11, No. 1, pp. 48-53.

14. Matalas, N. C., 1963, Probability Distribution of Low Flows, USGS Professional Paper 434-A.

15. Matalas, N. C. and J. R. Wallis, 1973, Eureka! It fits a Pearson Type III Distribution, Wat. Res. Res., Vol. 9, No. 2, pp. 281-289.

16. Nash, J. E. and J. Amorocho, 1966, The Accuracy of the Prediction of Floods of High Return Period, Wat. Res. Res., Vol. 2, No. 2, pp. 191-198.

17. Pentland, R. L. and D. R. Cuthbert, 1971, Operational Hydrology for Ungauged Streams by the Grid Square Technique, Wat. Res. Res., Vol. 7, No. 2, pp. 283-291.

18. Pentland, R. L. and D. R. Cuthbert, 1973, Multisite Daily Flow Generator, Wat. Res. Res., Vol. 9, No. 2, pp. 470-473.

19. Reich, B. M., 1973, Log Pearson Type III and Gumbel Analysis of Floods, Proc. Second International Symposium in Hydrology, Floods and Droughts, Water Resources Publications, Fort Collins, Colorado, pp. 291-303.

20. Riggs, H. C., 1968, Frequency Curves, Chapter A2, Book 4, in Techniques of Water Resources Investigations of the USGS, U. S. Government Printing Office, Washington, D. C.

21. Santos, A., 1970, The Statistical Treatment of Floods Flows, Water Power, Vol. 22, No. 2, pp. 63-67.

22. Solomon, S. I., T. P. Denouvilliez, C. Cadou and E. J. Chart, 1968, The Use of a Square Grid System for Computer Estimation of Precipitation, Temperature and Runoff in a Sparsely Gauged Area, Wat. Res. Res., Vol. 4, No. 5, pp. 919-930.

23. Solomon, S. I. and A. S. Qureshi, 1972, Application of a Parametric Model for Estimating Snow Accumulation and Flow Forecasting, Proc. IHD/UNESCO/WMO Symposia on the Role of Snow and Ice in Hydrology, Banff.

24. Spence, E. S., 1973, Theoretical Frequency Distributions for the Analysis of Plains Streamflow, Can. J. Earth Sciences, Vol. 10, pp. 130-139.

25. Wood, E. F. and I. Rodriguez-Iturbe, 1975, Bayesian Inference and Decision Making for Extreme Hydrologic Events, Wat. Res. Res., Vol. 11, No. 4, pp. 533-542.

27. Yevjevich, V., 1964, Fluctuations of Wet and Dry Years, Part II, Analysis by Serial Correlation, Hydrology Paper No. 4, Colorado State University, Fort Collins, Colorado.

27. Yevjevich, V., 1968, Misconceptions in Hydrology and Their Consequences, Wat. Res. Res., Vol. 4, No. 2, pp. 225-232.

28. Yevjevich, V., 1971, Stochasticity in Geophysical and Hydrological Time Series, Nordic Hydrology, Vol. II, pp. 217-242.

29. Zelenhasic, E., 1970, Theoretical Probability Distributions For Flood Flows, Hydrology Paper No. 42, Colorado State University, Fort Collins, Colorado.